Wolfgang Jendsch

Aerial Firefighting

4880 Lower Valley Road, Atglen, Pennsylvania 19310

Other Schiffer Books on Related Subjects

Innovations in Aircraft Construction. Hans Redemann. ISBN: 0887403387. $29.95.
Lockheed C-130 Hercules and Its Variants. Chris Reed. ISBN: 0764307223. $29.95.
One-of-a-Kind Research Aircraft: A History of In-Flight Simulators, Testbeds, & Prototypes. Steve Markman, Bill Holder. ISBN: 0887407978. $45.00.
Planes of the Presidents: An Illustrated History of Air Force One. Bill Holder. ISBN: 0764311875. $19.95.
USAF Plus Fifteen: A Photo History 1947-1962. David W. Menard. ISBN: 0887404839. $24.95.

Copyright © 2008 by Schiffer Publishing, Ltd.
Library of Congress Control Number: 2008930572

All rights reserved. No part of this work may be reproduced or used in any form or by any means—graphic, electronic, or mechanical, including photocopying or information storage and retrieval systems—without written permission from the publisher.

The scanning, uploading and distribution of this book or any part thereof via the Internet or via any other means without the permission of the publisher is illegal and punishable by law. Please purchase only authorized editions and do not participate in or encourage the electronic piracy of copyrighted materials.

"Schiffer," "Schiffer Publishing Ltd. & Design," and the "Design of pen and ink well" are registered trademarks of Schiffer Publishing Ltd.

Translated by Dr. Edward Force, Central Connecticut State University

Cover Design: Bruce Waters
Designed by Mark David Bowyer
Type set in News Gothic BT

ISBN: 978-0-7643-3068-1
Printed in China

Photo Credits: see page 348

Schiffer Books are available at special discounts for bulk purchases for sales promotions or premiums. Special editions, including personalized covers, corporate imprints, and excerpts can be created in large quantities for special needs. For more information contact the publisher:

Published by Schiffer Publishing Ltd.
4880 Lower Valley Road
Atglen, PA 19310
Phone: (610) 593-1777; Fax: (610) 593-2002
E-mail: Info@schifferbooks.com

For the largest selection of fine reference books on this and related subjects, please visit our web site at **www.schifferbooks.com**
We are always looking for people to write books on new and related subjects. If you have an idea for a book please contact us at the above address.

This book may be purchased from the publisher.
Include $5.00 for shipping.
Please try your bookstore first.
You may write for a free catalog.

In Europe, Schiffer books are distributed by
Bushwood Books
6 Marksbury Ave.
Kew Gardens
Surrey TW9 4JF England
Phone: 44 (0) 20 8392-8585; Fax: 44 (0) 20 8392-9876
E-mail: info@bushwoodbooks.co.uk
Website: www.bushwoodbooks.co.uk
Free postage in the U.K., Europe; air mail at cost.

Foreword

For many years there has been a fascination, and not only among laymen, with the themes of "aircraft" and "fire" or "firefighting." When these two related subjects are united into "firefighting from the air," then this subject awakens very special, focused attention among both firefighting specialists and "firefighting laymen."

In the summer months, the media also see to it that the often-spectacular fires in the forests of Australia, Russia, and the USA or the woodlands of Europe regularly become known to the public and thus remain timely in peoples' minds.

But what really takes place at the fire airbases, in the airplanes and helicopters, or at the scenes of action in and over the burning woods and wilderness, what special technology and tactics firefighters and foresters use most effectively, will be made clear in this book in the realm of forest and surface firefighting from the air.

The contents concentrate on the factual basis for "aerial firefighting" in the USA. The concepts that prevail there lead the way for the whole world and serve as a model for firefighting in Europe as well as in many other countries around the world. I myself have had opportunities over the years to learn detailed information about the technical and tactical sides of forest firefighting from the air, to observe airtankers and firefighting helicopters in action, and to take part in what happens in the world of aerial firefighting.

This book is neither able nor intended to be a reference work or catalog of aircraft and aircraft data—but it will try to portray a special kind of firefighting in generally understood form and based on personal experience, which is not usually accessible to the public.

Yet in this book interested readers will find a variety of aircraft of all types and functions that are used in the USA, Europe, and the world to fight forest and surface fires from the air. The palette ranges from the large airtanker through firefighting helicopters and transport planes for smokejumpers to the special control and command planes, especially the various types of American aircraft used to fight forest fires, which will be fairly completely depicted in this book.

Technical data, as well as tips for tactical use, excerpts from my own experiences on the scene, plus events in the field for this type of firefighting complete the informative contents.

One of my other books to be published by the Motorbuch Verlag—likewise concentrated on the USA—will dwell at length and in similar form with the landbound vehicles used for forest and surface firefighting.

I wish you an exciting time reading this book.

Wolfgang Jendsch
July 2007

Contents

International Firefighting Aircraft—Europe, World

The Future of Aerial Firefighting

Information and Tips

Aerial Firefighting

The History of Firefighting with Aircraft

The History of Firefighting with Aircraft

The history of firefighting from the air—aerial firefighting—in the United States of America and other parts of the world is long, highly interesting, and important in view of the common practical, factual knowledge and experience of the onetime "flight pioneers" and forest fire specialists. The basic concept of the first "fire managers" and "flight pioneers" in making use of the so-called "agricultural aircraft" (agrarian planes) proved from a present-day standpoint to be brilliant. The almost worldwide spread of concepts from the American western states of California, Oregon, and Washington makes clear that the fire managers of the early days were right in their concept of offering practical support from the air to the ground-based firefighters. Thus the development of aerial firefighting eventually developed in four major steps:

- The supervision and control of large areas of woodland and wildland from the air,
- The development of agricultural aviation into firefighting aviation,
- The idea of aerial firefighting, up to the founding of an official airtanker squadron for fighting forest fires from the air,
- The national and worldwide developments of aircraft conceived for woodland and wildland firefighting.

The following chapters offer a brief overview of these developments of aircraft concepts, and also describe the technical development of aircraft used in fighting forest and surface fires.

Supervision and Control

The history of aviation for the observation of wooded areas and fighting of forest fires in the USA began in the first years of the twentieth century. In 1919, for the first time, the U.S. Forest Service (USFS) introduced several small airplanes to make patrol and control flights over extensive, unreachable wooden areas of the western states of California, Oregon, and Washington that could be endangered by forest fires.

Available at that time were, among others, the light Curtiss JN-4D "Jenny" biplane.

In 1930, for the first time, a Ford Trimotor of the Stout Pullman 3-AT type was used, which could drop water from a wooden container while in flight. The flyers gave the nickname of "Tin Goose" to this somewhat "stilted"-acting plane, covered in waved sheet metal, with three exposed motors.

Later in the thirties, airplanes of this type, characterized unanimously by pilots as "good-natured," were used more often to fly additionally needed materials to the land-bound firefighters and drop them there directly or by parachute.

At the beginning of the forties, the first "smoke-jumpers" (special airborne forest firefighting units) were landed by parachute in order to make the initial firefighting attacks in woodlands and wildlands as well as set up fire lines in the areas around forest fires. Further experiments in dropping water on fires in containers or balloons from airplanes were soon given up for safety reasons. In the end, "shots" of this kind proved to be more dangerous to firefighters on the ground than the fire itself.

After World War II—Agricultural Aviation

After the end of World War II, many former military pilots looked for work with American charter flight companies that were active in the field of agriculture. In the Central Valley area of northern California in particular, with its large, fertile farms, many of these ex-military pilots found they could earn a good income spraying fields and plantings with insecticides and fertilizers.

The number of firms that were active in this realm increased considerably in the early fifties. In many cases, a pilot bought himself an airplane and founded his own agricultural flying company. Inventive owners equipped such planes as the Boeing N3N, Curtis (NAF) TS-2/3 or NAS N3N biplanes with additional technical spraying devices to spread fertilizer or insecticide over the extensive fields.

With the further development of agricultural aviation, the flying of materials and extinguishers for forest firefighting was also developed by the U.S. Forest Service.

The Development of Agricultural Aviation

Because of the growth of agriculture, especially in the Central Valley (Sacramento Valley, San Joaquin Valley) of California, the history of American agricultural aviation began in that region after World War II. Many names of firms and pilots are now legendary, such as Floyd Nolta, the founder of the Willows Flying Service in Willows, northern California.

Technical Data of the Stout Pullman 3-AT	
Year built	1928
Motor	3 Pratt & Whitney R-895 radial engines, 450 HP each
Crew	Pilot, co-pilot, spotter
Dimensions	15.10 meters long, 4.20 meters high
Wingspan	23.70 meters
Range	ca. 880 km
Speed	100 to 180 kph
Altitude	5,300 meters maximum

Technical data for the Ford tri-motor high-wing plane. From this plane, jokingly called the "Tin Goose," water was dumped from a wooden barrel.

Nolta also flew the lead plane in the dramatic war film *Thirty Seconds Over Tokyo*. The film portrays in detail the Doolittle surprise attack of the U.S. Air Force on Tokyo, under the command of Lieutenant Colonel James "Jimmy" Doolittle, on April 18, 1942. During the filming, Nolta was persuaded to fly under the San Francisco Bay Bridge in this twin-engine Mitchell B-25 bomber.

The film role played by Floyd Nolta is likewise typical of the role of agricultural pilots at that time. They were regarded as daring, unafraid, and venturesome, a kind of "heroism" that was and is also ascribed to the later pilots of the airtankers. After the war, Nolta reorganized his Willows Flying Service business along with his brothers Dale and Vance.

Other firms of this type in California—founded and led by WW II veterans, included Varney Air Industries (Raymond Varney), Sherwood Flying Service (Lee Sherwood), and Hendrickson Air Service (Harold Hendrickson).

A particular example in the development from agricultural aviation to modern-day airtanker production is offered by the Air Tractor Inc. firm of Olney, Texas. Its founder, Leland Snow, got into the technical development of the postwar era and produced agricultural aircraft in the fifties.

From 1951 on, he designed and developed the Snow-1 (S-1), the SR-2 Thrush in the sixties, and as of 1972 the first Air Tractor (AT-300). Later came numerous further types and versions. All of these planes, but especially the Air Tractors, were not only very successful in the growing agricultural aviation of the time, but also to this day in aerial firefighting.

See also:
Firefighting Planes/ air tankers in the USA—Single Engine airtanker (SEAT)

The Idea of Aerial Firefighting

Early in the forest fire season of 1955, American fire manager John Ely of the U.S. Forest Service (USFS) discussed his idea for using agricultural airplanes to fight forest and wildland fires, with various forest specialists and agricultural pilots in the region. At a meeting in Redding, California, he received authorization from Forest Supervisor Bob Dasmann to continue his air-tanker project.

It was also John Ely who drew the experienced veteran agricultural pilot Floyd Nolta into the discussion and asked him to dump water on a forest fire from an airplane. Nolta was known among pilots as not only daring and venturesome, but also as a man with vision and extraordinary energy. Within a short time he put Ely's idea into practice by cutting a hole in the bottom of his Boeing Stearman biplane and attaching a hinged flap that could be opened by a cable. The plane, filled with water, thus became the first specially prepared firefighting airtanker. This new development was tested at an airport in California on July 23, 1955. Grass was set afire beside the runway, Floyd Nolta flew over the fire, released the extinguishing water—and put out the fire within a few seconds!

Technical Data, Boeing Stearman Caydet	
Year built	1936
Motor	Pratt & Whitney Continental R-670-5 radial (450 HP)
Crew	Pilot
Dimensions	length 7.60 meters, height 2.80 meters
Wingspan	9.80 meters
Range	ca. 810 km
Speed	ca. 200 kph
Altitude	11,200 feet (ca. 3.4 km)

Technical data of the first American airtanker, from Vance Nolta.

Things got serious at the Mendenhall Fire on August 13, 1955. Floyd Nolta's brother, Vance, was called by the Fire Control Office of the Mendocino National Forest (west of Willows, northwest of Sacramento), after a forest fire had broken out there. After the success extinguishing the "grass fire" at the airport in Willows, the Willows Flying Service had fitted a Boeing Stearman 75 Caydet agricultural biplane with a 170-gallon tank, which was now to see its first service in a real forest fire.

By dropping his load of water, Vance Nolta was able to extinguish the fire on the west side of Bald Mountain in the Covelo district. After each drop he flew back to Gravelly Valley and had his tank filled by the water tender of the Ukiah Pine Fire Department. Nolta dropped his load of water on the Mendenhall fire six times, always in cooperation with the fire crew on the ground. The fire was extinguished at last. This was a success and an achievement that had not been expected at first from the crew and those in command, yet, the event was highly praised afterward.

Vance Nolta's plane, N75081, was registered as the first airtanker in the history of aviation to fight forest and wildland fires. This event was likewise the "birthday of the American air-tanker program, and its 50th anniversary was celebrated in Willows, California, on Saturday, August 13, 2005.

Aerial Firefighting Becomes Official—The Airtanker Squadron

In the following years, many of the available agricultural planes were modified for firefighting use. With their seven aircraft, the First Operational airtanker Squadron in the United States began in 1956, and in the following fire seasons they set out from the Willows airfield to see service all over California.

Very soon the availability of the airtanker Squadron became known to the forest fire units of the U.S. Forest Service (USFS) and the California Department of Forestry and Fire

Operational Airtanker Squadron		
as of 1956		
Number	**Plane Type**	**Pilot**
	Squad Chief	Floyd Nolta
75081 (1)	Boeing Stearman 75 Caydet	Vance Nolta
2	Naval Aircraft Factory (NAS) N3N-3	Dale Nolta
3	Naval Aircraft Factory (NAS) N3N-3	Harold Hendrickson
4	Naval Aircraft Factory (NAS) N3N-3	L.H. McDurley
5	Naval Aircraft Factory (NAS) N3N-3	Ray Varney
6	Naval Aircraft Factory (NAS) N3N-3	Warren Bullock
7	Naval Aircraft Factory (NAS) N3N-3	Frank Prentice
as of 1957		
8	Naval Aircraft Factory (NAS) N3N-3	Frank Michaud
9	Naval Aircraft Factory (NAS) N3N-3	Gene Ellan
10	Naval Aircraft Factory (NAS) N3N-3	Neal Wade
11	Naval Aircraft Factory (NAS) N3N-3	George Jess

The table lists the first American Airtanker Flight Season (Air Tanker Squadron) for combating forest and surface fires from the air (Aerial Firefighting).

Protection (CDF). In Willows, a command post was set up from which the airtankers could be called by telephone emergency number 80. In the first month after its establishment, the airtanker Squadron's planes saw action in twelve forest fires in California.

In the next year, 1957, the firefighting unit in Willows was equipped with additional planes and pilots. In the ensuing years, the light biplanes, and several Grumman TBM Avengers that had been added later but had likewise proved to be too light, were replaced by Consolidated PBY Catalina models—a flying boat used by the U.S. Navy for sea reconnaissance, by Grumman F7F Tigercats, and other large land-based airplanes.

From the 1960s onward, the U.S. Forest Service upgraded its fleet of airtankers with larger planes. In particular, ex-WW II military planes such as the Consolidated B-24 (a four-engine heavy bomber), Douglas A-26 (twin-engine light fighter plane), Douglas DC-6 (four-engine passenger plane), and Boeing B-17 (U.S. Air Force bomber) were rebuilt and fitted with 2500-gallon (ca. 9460-liter) tanks for retardant. The increased loads of firefighting materials proved to be much more effective for fighting forest and wildland fires than the comparatively small "drops" of the first-generation airtankers.

Retardant Instead of Water

In the practice of forest fire fighting, it was quickly recognized that dropping water from planes did not bring the hoped-for success on the ground. Depending on weather and vegetation conditions, and in large fires with high flames and great heat development, the water often did not even reach the ground. Thus, in 1956, natrium-calcium borate chemical salts were first added to the water to counteract the fast evaporation of the water after it was dumped. As a result, the firefighting planes came to be called "borate bombers."

Long since aeronautical history: The first Boeing B-17 ex-Air Force bombers were rebuilt for forest fire fighting.

Later the borate was replaced by Bentonite, a mineral material from clay-holding rock (which contains the clay mineral montmorillonite), which is much lighter and, together with water, swells up and forms a gel-like substance.

After long-term tests based on experience gained in practical forest fire fighting from the air, the California fire managers finally decided in favor of "Firetrol," a gel-like cooling and extinguishing substance consisting mainly of phosphates, sulphates, and iron oxide dissolved in water, plus other extinguishing and limiting substances.

Firetrol and similar products, made by other manufacturers, are still used today in firefighting from the air.

The Development of CDF Aviation Management

The California Department of Forestry and Fire Protection (CDF) also took part in the development, spreading from California, of agricultural aviation into firefighting aviation.

During the 1954 to 1957 forest fire seasons, though, the organization went its own way and prepared smaller agricultural airplanes reequipped for forest and wildland firefighting at so-called "call when needed fire air bases."

In 1958 the CDF chartered the first air-tankers from private aviation firms that had already specialized in aerial firefighting.

In the following years, the CDF used former military and civilian aircraft that were converted for forest and wildland firefighting, among them the Twin Beech, Grumman AF-2 and Grumman F7F, Consolidated PBY, and Boeing B-17. In the early 1970s, the CDF tanker fleet already consisted of 14 TBM Avengers, five F7F, one PBY, and one B-17.

A rare photo of a Consolidated PBY4 (Airtanker 126) equipped with a drag chute, which enabled "short landings" on limited-size fire air bases.

From the earliest days of the CDF air-tanker fleet: An S-2A of the California Department of Forestry and Fire Protection (CDF) in green-gray colors is seen dropping water.

Boeing B-17 airtanker 71 (No. N5233V).

Table: Data on the Development of CDF aviation

CDF Aviation			
as of 1958			
Number	**Aircraft Type**		**Type**
3	Naval Aircraft Factory (NAS) N3N-3	Airtanker	Contract/Conversion
4	Boeing Stearman 75 Caydet	Airtanker	Contract/Conversion
4	TBM Avenger	Airtanker	Contract/Conversion
as of 1965			
	Bell 47	Helicopter	Contract
	Fairchild-Hiller FH 1100	Helicopter	Contract
6	Bell 206 Jet Ranger Helitack	Helicopter	Contract
	Aerospatiale Alouette III	Helicopter	Contract
as of 1970			
	Twin Beech	Airtanker	Contract/Conversion
	Grumman AF-2	Airtanker	Contract/Conversion
1	Boeing B-17	Airtanker	Contract/Conversion
1	Consolidated PBY	Airtanker	Contract/Conversion
5	Grumman F7F	Airtanker	Contract/Conversion
14	TBM Avenger	Airtanker	Contract/Conversion
as of 1974			
12	Grumman S-2A	Airtanker	Contract/Conversion
55	Grumman S-2A	Airtanker	Contract/Conversion
20	Cessna O-2 Skymaster	Air Tactical Aircraft	CDF (exUSAF)
1978			
3	Bell 205 (mil. Bell UH-1H)	Helicopter	CDF (exUSAF)
1981			
12	Bell UH-1 F Hueys	Helicopter	CDF (exUSAF)
6	Bell UH-1 F Hueys	Helicopter	CDF (exUSAF)
as of 1984			
	Fixed tanks introduced for helitack	Helicopter	CDF
as of 1987			
	Grumman S-2T	Airtanker	CDF prototype
	Grumman S-2T	Airtanker	CDF exchange S-2A
26	Grumman S-2T E/G	Airtanker	CDF newly built
as of 1991			
	Bell UH-1H & 1F Super Huey	Helicopter	CDF modification
as of 1993			
13	Rockwell OV-10A Bronco	Air Tactical Aircraft	CDF (exUSAF)

Former AT-136 of the Fairchild C-119 type, long since mustered out, is seen on a training flight over Twin Falls, California. Because of their design, these planes were jokingly called "flying boxcars."

In the mid-seventies the CDF chartered twelve Grumman S-2A airtankers and numerous single-engine Cessna 182 and Cessna 210 planes (air tactical aircraft) of four contract businesses: Aero-Union Corporation, Sis-Q Flying Service, TBM Inc., and Hemet Valley Flying.

In the practice of forest and wildland firefighting, though, the Cessnas proved to be too slow and too unsafe for action.

For these reasons, CDF Senior Air Operation Officer Cotton Mason, in 1974, chose twenty twin-engine Cessna 337 O-2 Skymasters from a supply of forty U.S. Air Force planes, and had the former Vietnam control planes transported to the airtanker base in Fresno.

In 1976 the O-2 was put into service and proved to be a success by CDF standards. The planes were used as leadplanes and command planes.

This Lockheed PV-2 Harpoon (Type III-AT-39, No. N7080C), of Hirth airtankers in Buffalo, Wyoming, is already an oldtimer too. It was a technical derivative of the previous PV-1 Ventura and the predecessor of the P2V Neptune airtanker. After World War II, in 1948, some former Navy bombers were taken over by charter companies, which used the planes to fight forest fires. The plane is 51 feet (some 15.5 meters) long, with a wingspan of 75 feet (ca. 22.90 meters); its top speed is 282 mph (ca. 454 kph). Hirth airtankers operated airtankers 37, 38, and 39, which were each fitted with an 800-gallon (ca. 3028-liter) retardant tank.

The CDF completed another partial aviation program in 1993. Sixteen two-seat turboprop Rockwell OV-10A Bronco planes were taken over from the U.S. Navy and intended to replace the O-2. In all, thirteen of these planes are now in service with the CDF.

A new generation of airtankers was introduced in 1970 with the Grumman S-2 Tracker, a former submarine fighter of the U.S. Navy. In general, the S-2 replaced the TBM Avenger. Rebuilt (tanks installed) and later modified, the CDF planes are still used today by the Mobile Equipment facility in Davis, California, and the Hemet Valley Flying Service (equipment) in Hemet, California.

The S-2 fleet of the CDF already included ten planes in 1973; another twelve planes were added in 1974, and five more in 1975. In the mid-nineties, the S-2A airtankers (two turbines, each with 1500 HP, 800-gallon tank) were replaced by faster and more modern S-2T airtankers (two turbines, each with 1650 HP, 1200-gallon tank). In all, 23 S-2T planes especially equipped for forest and flatland firefighting were built by Marsh Aviation in Arizona and put into service by the CDF. At the end of 2002 another 13 new planes were added, seven more at the end of 2004, and three in June 2005.

Along with the airtankers, the CDF has also used helicopters since the mid-sixties to patrol forests and wildlands and fight forest fires. Until the seventies, for example, Bell 47 (Bell H-13 or Bell HTL in the military), Fairchild-Hiller FH 1100, Bell 206 Jet Ranger, and Aerospatiale Alouette II had been chartered from private contractors.

After several accidents, the CDF decided to acquire the department's own helicopters. As of 1981 the first twelve Bell UH-1F were put into service, soon to be replaced by the larger Bell UH-1H. This type of helicopter, modified to meet CDF needs, was known as "Super Huey".

In 1960 the CDF also set up six eleven-man Helitack Crews and stationed these new special units with their Bell 206 Jet Ranger helicopters at their own helitack bases in the early seventies.

In the expansion of the helicopter fleet, three medium-large Bell 205 helicopters were obtained in 1978, one of each being stationed at the headquarters of the Mendocino Rangers in Howard Forest, at the Hemet-Ryan Airfield, and at the headquarters of the San Diego Rangers in Monte Vista.

A further program, beginning in 1981, included the acquiring of, at first, twelve Bell UH-1F Huey helicopters and, later, six more of them.

All the helitack helicopters were used for firefighting with Canadian "Bambi Buckets" (324 gallons, ca. 1300 liters). In the mid-eighties, the first CDF helicopters were equipped with fixed tanks on their bottoms. Such a system offered greater safety, especially in inhabited areas—a safety that dropping a Bambi Bucket could not offer.

Amphibian airplanes (scoopers) are impressive and fascinating with their "bullish" structural style.

A Completely Different Concept—Amphibian Airtankers

A completely different concept in forest and flatland firefighting was based on the use of amphibian planes. Some of the many military and civilian planes built at that time, which could take off and land on both land and water, were used in converted form for aerial firefighting. Besides the Consolidated PBY Catalina/Canso, a former reconnaissance plane of the U.S. Navy, the Canadian JRM-2 Mars, an airtanker based on the U.S. Navy's Martin PBM Mariner flying boat, and the Russian Beriyev B-40 Albatross, a civilian rescue flying boat, the Canadair Group of Canada, a branch of Bombardier Aerospace, developed special firefighting planes since 1963. They were able to take water into their 5000-liter tanks by landing on the water's surface. Bombardiers and the so-called Canadair CL-215, CL-315T, and CL-415 are to date the only firefighting planes in the world designed especially for forest and flatland firefighting.

See also:
Firefighting Planes International—Water Bombers/Canadair Planes

The AT 802 Firebos can also be counted among the modern amphibian firefighting planes. This plane, built by Air Tractor of Texas and fitted with pontoons, can take on either water or retardant on land or with a scooper when landing on the water.

Amphibian airtankers are used today mainly in the Mediterranean countries of southern Europe (including France, Spain, Croatia, Greece, Italy) and in Canada. American fire managers have used Canadair planes only in exceptional cases or for testing. Only three Canadair CL-215 are stationed in the USA, one on North Carolina and two in Minnesota.

See also:
Firefighting Planes International—International Concepts

Amphibian planes of the Beriyev A-40 Albatross, Beriyev 12 Chaika, and Beriyev A-200, are used to fight forest fires in Russia. For economic as well as tactical reasons, amphibian planes are used primarily there, along with special firefighting planes such as the Ilyushin 76P. Along with former military planes rebuilt for firefighting purposes, modern airtankers have also been developed there, based on the multipurpose civilian Beriyev A-200 amphibian plane.

Worldwide Models

In the USA, fighting forest and flatland fires, and particularly aerial firefighting, have been regarded as specialized models since the beginning of their development. The technology and tactics of American firefighting planes have been adopted by many nations and prevail today, sometimes in modified form suited to regional needs.

A selection of the firefighting airplanes and helicopters used internationally to fight forest and flatland fires will be described and/or depicted in the following chapters.

Notes

In this and all the following contributions of this book, American technical terms (such as "air tanker" and "fire air base") are used. This has proved to be practical, not only because the contents focus extensively on American aerial firefighting, but also because the English language is widely used internationally.

As far as has been necessary for factual understanding, German terminology was also used in the original German edition.

An extensive listing of the essential specialist terms from American aerial and wildland firefighting (see pp. 344-345) will make understanding these concepts easier.

32
32

25
N925AU

Air Tankers in the USA

Technology and Tactics of Airtankers in Fighting Forest Fires

By numbers, airtankers constitute the largest part of all the tactical aircraft used for forest fire fighting worldwide. In a tactical sense too, they play the most important role in fighting forest fires from the air, especially in the USA and Canada.

See also:
The Future of Aerial Firefighting: Airtanker Crashes in Forest Fires, the end of the large airtanker in the USA.

Firefighting aircraft are defined here as either airplanes/aircraft that use either fixed wings or helicopter blades in flight. Amphibian planes with appropriate firefighting equipment also qualify a firefighting aircraft.

Firefighting aircraft are used for firefighting in general and forest or flatland firefighting in particular, either by their conceptual design or through their military or civilian conversion or equipping with either rigidly attached or flexible containers (buckets).

In the following section, and in reference to the American system of technology and tactics of aerial forest and flatland firefighting, the concept of the firefighting plane is assigned to fixed-wing aircraft. The corresponding concept used in the USA and partially in other English-speaking countries is "airtanker"—as opposed to the fire helicopters or helitankers that will be portrayed in a special second section.

Ex-Military Airplanes

Most airtankers and other aircraft were not developed originally for use in fighting forest and flatland fires. In the USA, primarily ex-military planes (freight and transport planes, bombers and patrol planes) which were converted for forest firefighting by specialized charter firms (contractors) and fire-protection agencies are used. These are largely older aircraft of World War II vintage, which have meanwhile been questioned in terms of safety by the National Transportation and Safety Board (NTSB) and the large national forest and forest fire organizations (U.S. Forest Service: USFS, Bureau of Land Management: BLM) and, since April 2004, retired for the most part.

Former military airplanes are left extensively in their original technical and visible structural condition when converted to airtankers by the large forest fire contractors (charter companies). In general, though, one or more tanks are installed firmly in a prescribed size order, which can discharge their contents through two to eight (depending on the type) hatches mounted on the bottom of the fuselage. Mars airtankers have 26 such hatches; Boeing KC-97 planes have 16.

Electric or electronic-hydraulic programming and releasing apparatus allow the pilot to drop the extinguishing materials from the cockpit.

Such tanks are not firmly installed in military airtankers of the U.S. National Guard or the U.S. Air Force, which are equipped with the mobile Modular Airborne Fire Fighting System (MAFFS, or more recently AFFS).

Airtankers are also equipped with the Global Positioning System (GPS), with targeting systems for the most exact dropping of extinguishing materials, and with radios. In a few cases there is a heat-protection system installed in the cockpit.

ICS-Types

Along with the general size designation of "large/heavy airtankers," "standard/medium airtankers," and "small/light airtankers," all airtankers in the USA are divided into "types." These "types" correspond to the prescriptions of the American Incident Command System (ICS) and, as a rule, are classified according to the quantity of material they carry. These aircraft are divided into Type I, II, III, and IV (see the table below):

Types of Airtankers (USA) According to Incident Command System (ICS)		
	Quantity of Extinguishing Materials	
Types	**Gallons**	**Liters**
I	more than 3000	more than 11,345
II	1800 to 2999	6813 to 11,345
III	600 to 1799	2271 to 6812
IV	100 to 599	379 to 2270

Airtanker Types I to IV, classified by the American Incident Command System (ICS). The table lists the material capacities of the firefighting planes.

A classification of firefighting aircraft into ICS or general classes is scarcely done, if at all, outside the USA, except in Canada. At best, a classification as large or heavy, or small or light airtankers is used.

Tactically, firefighting airplanes (airtankers) in the USA are linked to a close cooperative system between aerial firefighting and ground firefighting. A firefighting operation is, as a rule, led by an Incident Commander Ground and an Incident Commander Air. Aircraft (airtankers, helicopters, and other aircraft) are generally under the command of an air commander (Incident Commander Air), who flies his command plane over the whole field of operations.

See also:
Firefighting Planes/ Air Tankers in the USA: The Modular Airborne Fire Fighting System (MAFFS)

At the places where the water or retardant is dumped, non-local or military airtankers are led by lead planes—small and maneuverable planes that serve as leaders or "follow me" planes for the pilots, who generally do not know the local area. Lead planes direct while in close radio communication with the Air Commander and the Incident Commander Ground, and in special cases also directly with a firefighting unit (fire crew, ground crew, engine crew) on the ground.

Airtanker Contractors

In the USA, airtankers are chartered for a forest fire season by the national agencies (U.S. Forest Service or Bureau of Land Management) or, in special cases, by state forest and forest fire agencies from commercial charter firms specializing in forest and flatland firefighting from the air (aerial firefighting), by National Large Airtanker Services contracts. The contract conditions are based on detailed requirements provided by the agencies, which are to be followed precisely by the contractors. They concern the fire-protection technical equipment on the chartered aircraft as well as the personnel, the duration of activity at specified fire airbases (special airports for aircraft used for fighting forest and wildland firefighting), or technical warning systems.

See also:
Information and Tips: Manufacturers and Contractors of Firefighting Aircraft (table)

State Airtankers

The California Department of Forestry and Fire Protection (CDF), the state organization that is responsible for forest and forest fire supervision, is the only state agency with its own numerous Type III airtankers, while the North Carolina Division of Forest Resources (NCDFR) obtained a Canadair CL-215 in 1998 and the Minnesota Department of Natural Resources bought two Canadair CL-215 airtankers in 2001. The State of Minnesota is thus, along with the states of Michigan and Wisconsin and the Canadian provinces of Ontario and Manitoba, a member of the Great Lakes Forest Fire Compact (GLFFC). This organization uses additional Canadair airtankers in the region of its member states and provinces.

Technical Equipment of the Airtankers

The technical equipment of the firefighting airplanes varies much inside and outside the USA. The standard equipment includes various sizes of water or retardant tanks, arranged either as single or linked tanks. These tanks are emptied through varying numbers of hatches on the bottom of the plane's fuselage. According to the type and equipment of the machine, one to eight hatches are installed, and a few airtankers have up to 16. As a rule, these hatches can be operated in different ways, so that either one is opened alone, several (usually two) together, or all the hatches at the same time to dump the water or retardant. The tank hatches are operated either manually-hydraulically or electronically from the cockpit. Sometimes modern computer-operated tank supervision, guiding, and apportioning systems are included.

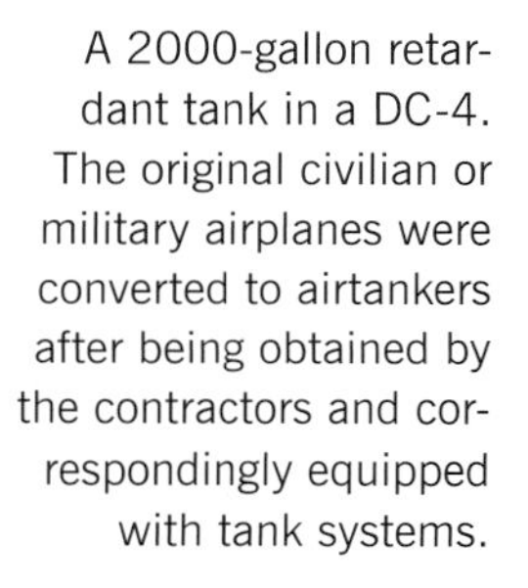

A 2000-gallon retardant tank in a DC-4. The original civilian or military airplanes were converted to airtankers after being obtained by the contractors and correspondingly equipped with tank systems.

Three of the six retardant-tank hatches of the P2V-7 are shown opened. The maximumn load of the Airtanker 08 is 2400 gallons (ca. 9080 liters).

As a rule, American airtankers are equipped with global positioning systems (GPS), so that in addition to the navigational instruments required for all aircraft, they have an additional instrument for a target-aimed approach onboard. Among the new audio-visual equipment of airtankers is, for example, the Airborne Bispectral Imager (Daedalus ABS) made by the American firm of Angiel Envirosafe, Inc., a graphic information system (GIS), which provides three-dimensional presentations of forest and flatland fires on an infra-red or ultra-violet basis, thus enabling targeted pinpoint drops from airtankers.

American airtankers, and usually firefighting aircraft of other countries as well, generally have two different radio communication systems onboard. One is the regular aircraft radio, which connects them with the responsible tower or the civilian air supervision; the other radio communicates with the responsible radio command posts of the towns, counties, or forest and forest fire authorities. Regional agencies for forest and flatland firefighting (Aerial Firefighting, Wildland Firefighting), special radio channels (air and ground units) are thus available. Military firefighting planes, airtankers (MAFFS, etc.), also have non-public military radio communication.

By no means do all airtankers have air conditioning and/or special heat protection for the cockpit. Often they have only aluminum covers for the cockpit windows, which—fitted only with small vision louvers—offer minimal protection from heat radiation in a flight directly over the fire.

Tactics of Aerial Firefighting

The following introduction to the special tactics of aerial firefighting – as practiced in the USA and adopted either in similar forms or on occasion in other lands – neither can nor will be a textbook on the subject of forest firefighting from the air. This is neither the purpose of this book nor the appropriate place as such a detailed recounting is outside the bounds of the intended presentation.

The basic extinguishing and attacking methods (basic attacks) from firefighting aircraft (airtankers) described below, which are adapted to the tactics of the ground crews, make clear the complexity of the tactical procedure and the required cooperation between air attack and ground attack units. Airtanker action is basically effective only in close cooperation with the ground crews. They require the accurately aimed, pinpoint drops of water and/or retardant from the aircraft, which are utilized tactically on the ground as a kind of fire line.

Airtanker drops—especially of retardant—which are usually dropped tactically right on the flanks, or in particular cases in front of a forest or flatland fire, are actually not a means of extinguishing the fire, but rather a means of cooling that supports the firefighting work on the ground, intended to reduce the intensity and spreading of the fire effectively on a long-term basis. This requires both punctual and intensive further work of the units on the ground, so as to be able to extinguish the remaining fire, burning nests, and glowing embers once and for all.

In the European countries in particular, and in most countries of the world, airtankers generally work only with water, and in certain cases also with foam additives. Here the emphasis is on the immediate (short-term) extinguishing effect of a drop. Depending on the structure, organization, and chosen tactics of the firefighting units on the ground, close cooperation works only at the start.

In addition, lead planes or command planes, such as are used in American aerial firefighting, are often not included in forest and flatland firefighting outside the USA. Directing and advising the airtankers then is usually done by radio, or by sight contact with the pilots.

Safety, Effectiveness, Efficiency (SEE)

"SEE—what you can do!"—this motto is vitally important to the crews of the airtankers when they go into action fighting a forest fire. "SEE what you can do!" also includes the initials of three important slogans for forest firefighting from the air, that should be, and are, followed in this order:

- **S = Safe**

  Safety—your own, that of the crew and that of the units on the ground—has the highest priority for firefighters in the air, and precedes all other measures;

- **E = Effective**

  Firefighting as fast and effective as possible is the goal of all efforts, not only of aerial firefighting. Effectiveness and success, though, are always subordinate to the safety of all measures.

- **E = Efficient**

  Questionable "virtues" such as "heroism" and "daring" are generally not asked for in American wildland firefighting, and certainly not in aerial firefighting. The personal achievement of the individual is thus less important than the safety and effectiveness of all measures.

According to this motto, tactics for pilots and crews of firefighting planes basically means "a series of tasks and operations, done to activate objectives."

Flanking—a tactic in which a forest or flatland fire is fought from the back along its sides.

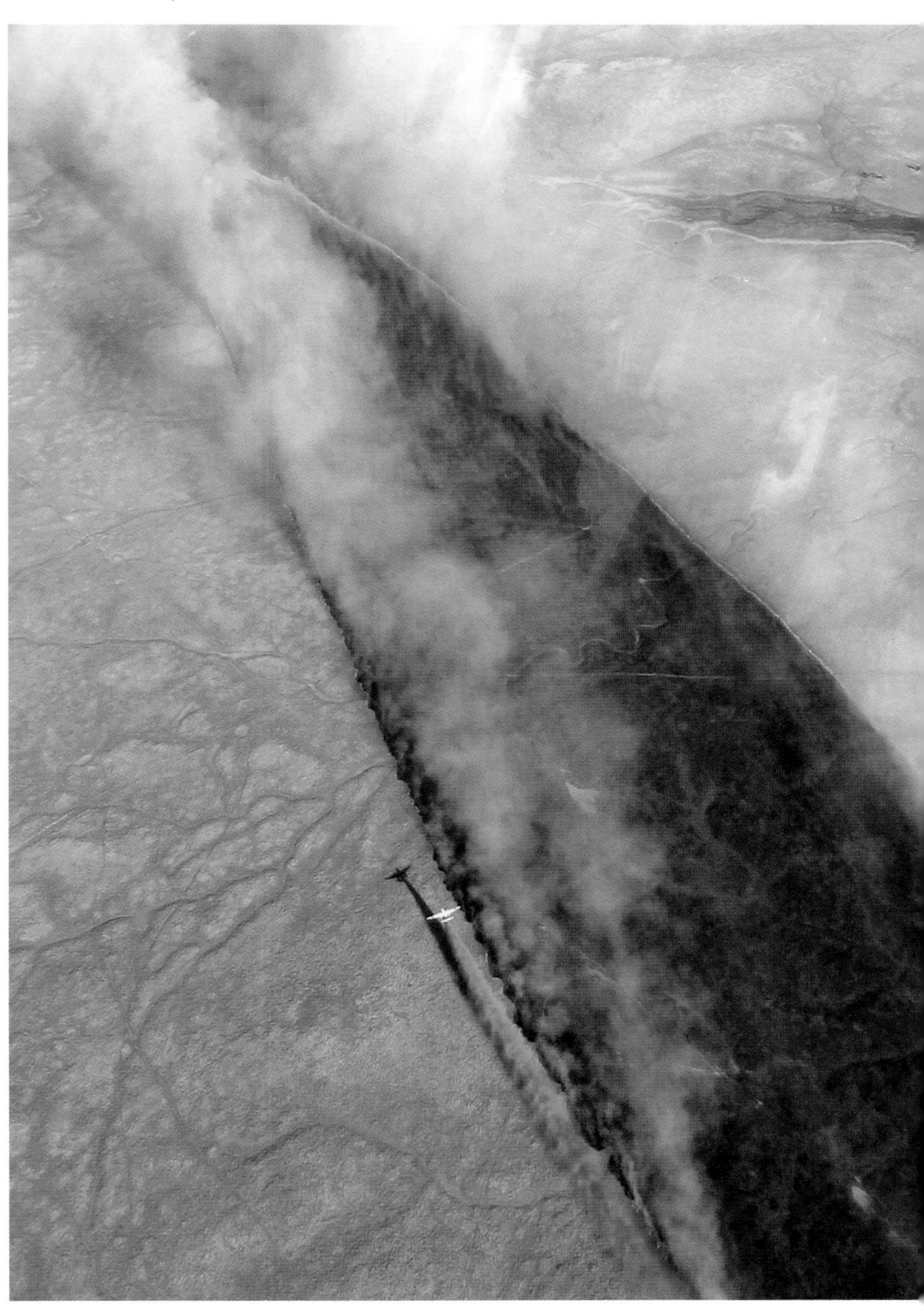

The graphic shows the separate phases of a standard run by an airtanker in a forest or surface fire. The planes generally fly in the direction of the wind, along one flank, to the head of the fire, and there turn in the opposite direction (turning base—base-turning final), to make the retardant drop on the other flank.

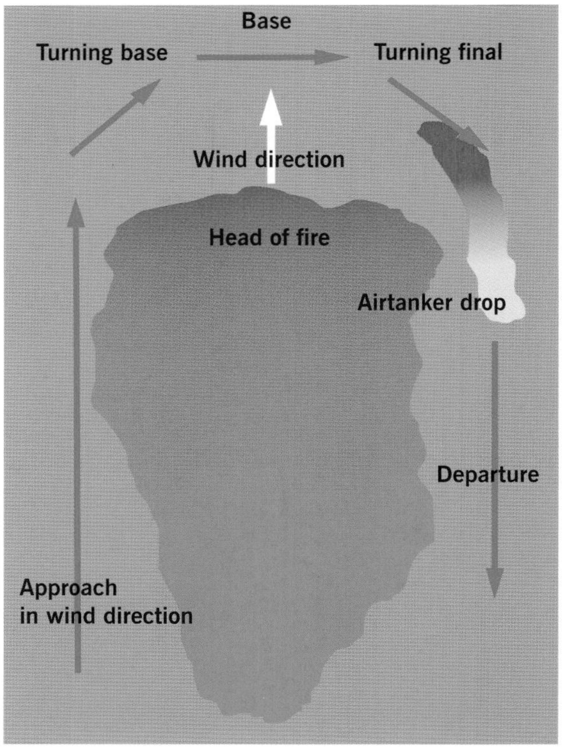

In tactical terms, there are three basic methods of extinguishing a fire used in forest and flatland firefighting in the USA:

- The Indirect Attack,
- The Direct Attack, and
- The Parallel Attack.

At first the concepts are simple to understand: the **indirect attack**, in which the fire, in terms of its direction of advance, is approached from the rear by forward-moving control and defensive measures, using natural terrain conditions to control the fire in front; the **direct attack,** in which the fire, again in terms of its direction of advance, is approached laterally, from the flanks; and the **parallel attack,** in which several tactical measures are carried out together in a narrow space.

The airtankers are subordinated to these basic, ground-linked attack methods, meaning that they suit their tactical measures (dropping of water and/or retardant) to the ground tactics. In the practice of aerial firefighting, though, these approaches are very much more complex and depend on certain prerequisites (such as weather conditions, vegetation, terrain, air and ground resources).

The most important tactical firefighting attack measures for airtankers, following the American aircraft patterns, will be described below in simple and understandable form.

See also:
Flanking, picture on page 29, graphics on page 31-1.

Flanking

"Flanking fire suppression" refers to an approach to the fire on one or both flanks (side areas). These extinguishing attacks are carried out in the direction in which the fire is spreading (wind direction) from a relatively inactive area or a tactically favorable and non-dangerous area (anchor point) in the back part of the fire. Flanking is a very effective tactic, especially with fast-moving fires.

Tandem Ahead

Tandem refers to a combined tactic and procedure by two or more different air or ground units (for example, airtanker and dozer), which carry out their firefighting activities immediately one after the other. "Tandem Ahead" is a forward-moving tactic (in the direction of the fire), in which the appropriate measures are applied as nearly simultaneously as possible.

A "tandem ahead" tactic applied by airtankers proves in practice to be especially helpful support to the ground crews, especially to the dozer crews who are setting up fire lines.

Tandem Behind

The "tandem behind" tactic is a support action independent in time but carried out in close cooperation, for example, between two engine crews (teams with firefighting vehicles) and airtankers. It is carried out particularly to hold already applied fire lines and to protect buildings.

Hook

Support tactics for ground units to hold already applied fire lines in flank areas of a fire.

Narrow VEE—Wide VEE

An airtanker tactic in which a thin, only slightly spreading fire wall or a small-surface fire is stopped or extinguished by narrow or wide wedge-shaped drops of extinguishing or cooling materials. Such a quick attack to the head of the fire is, on the one hand, regarded as aggressive, and on the other hand as a timesaving method. Essential to its successful completion, however, is equally fast action by the ground crews in setting up fire lines or water lines. In this procedure, additional airtanker runs can be needed after an indefinite time.

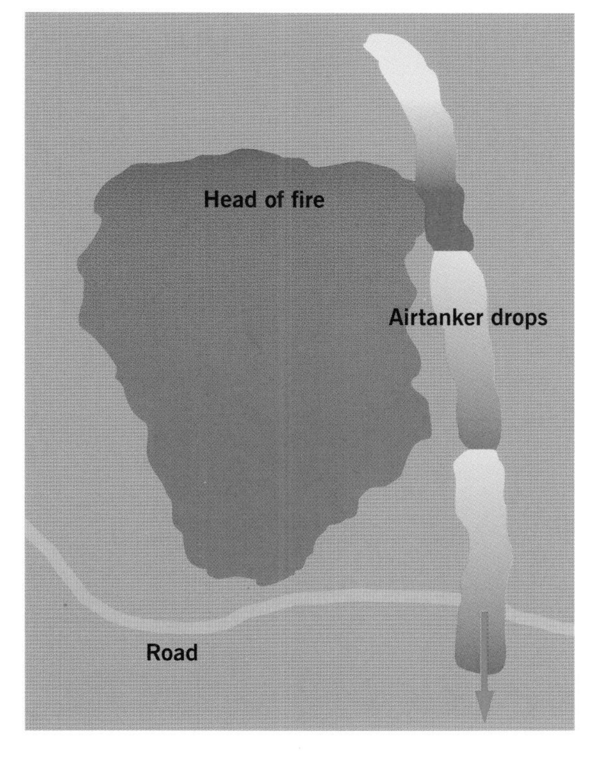

Flanking tactic: An extinguishing tactic for a fast-moving fire. The airtanker's attack is made from a so-called anchor point along the left or right flank in the opposite direction to the fire's movement.

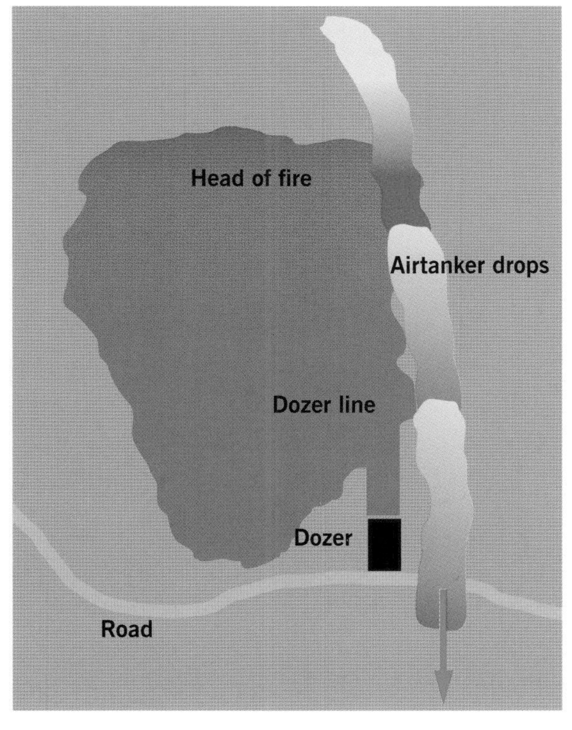

Tandem ahead tactic: An air-support measure for ground crews, especially for the use of bulldozers to set up fire lines. From behind and together toward the front on one of the flanks, the fire is attacked by airtankers, while the ground crews work their way forward along the retardant line.

Angle Tie-In

In the angle tie-in tactic, the work of the ground crews is supported by drops of extinguishing or cooling material, usually at right angles. Depending on the conditions at hand (for example, the direction in which the fire is spreading, natural barriers, etc.), airtanker drops can form a "tactical triangle" between the direction of the fire, natural barriers (such as roads or waterways), and retardant lines, which can then be worked on by ground crews (extinguished, burned out or mopped up). The angle tie-in tactic is an often-used cooperative method between air and ground units.

Spot Fire

A spot fire is a fire outside the main fire area, perhaps ignited by flying sparks, that increases the danger of further spreading. Such a small fire is usually covered or extinguished by several drops of retardant or water from two directions. Drops from two opposed directions provides better cover of spot fires in hard-to-reach areas as well.

Pre-Treat

Pre-treat is an advancing airtanker tactic in which fire lines are set up in anticipation of a spreading fire. Such fire lines are usually laid out by involving natural terrain barriers (such as ridges), to offer the ground crews standing ready in visible and protected areas an optimal starting point for control work.

Angle Out

An extended tactic of the "pre-treat," including further natural barriers (such as additional fire lines on canyon edges, besides the main fire line on a ridge. Such measures are carried out only in close cooperation with the ground crews and must be done appropriately early in tactical and technical terms.

Cooling Convection Current

This situation-dependent tactic could be described as cooling a spreading column of smoke and/or development of heat, in which airtanker drops are made specifically in the direction of spreading fire (see the graphic).

Spot Field Cover

A relatively new tactic in aerial firefighting is the spot field cover. It is used in forest fires that are typified by especially aggressive spreading and the initiation of spot fires. In spot field cover, several retardant lines are laid perpendicular to the fire's direction of spread and at varying intervals and distances. Thus even spot fires that are activated beyond one or more fire lines can be brought under control by limitation. A large number of airtankers would be needed simultaneously for spot field cover tactics.

One last graphic presentation should make clear again how airtanker action can depend on external conditions (such as wind direction) and how tactics that are as flexible as pos-

Tactics of an airtanker in fighting a surface fire near a main road, caused by a car fire. The fire, spreading to the left in the picture, is stopped by intersecting drops (Narrow Vee).

sible can lead to safe, effective, and efficient (SEE) success.

In the course of fighting a forest fire, it may become necessary for originally agreed-upon firefighting tactics (such as flanking as seen on the left side of the graphic) to be extended, added to or evaluated anew because of a change in wind direction resulting in a change in a fire's spreading direction, or an intensifying danger to previously unthreatened buildings. The graphic indicates such a case, in which the main fire plus small spot fires suddenly move in the direction of a building. The previous flanking action of an airtanker thus becomes ineffective and must be replaced by a modified angle tie-in tactic (at the right side of the graphic).

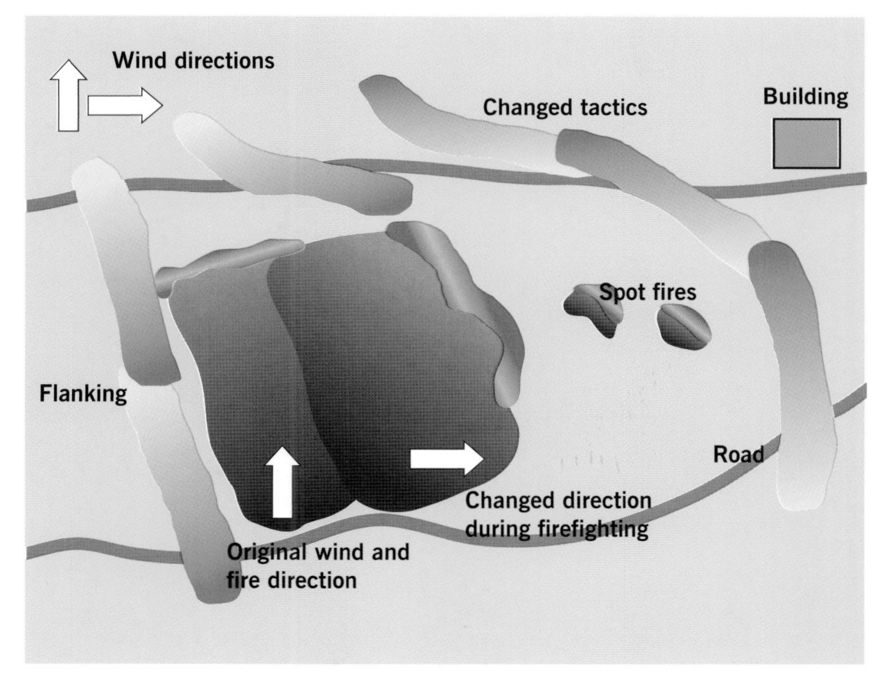

The graphic shows extinguishing tactics changed to deal with weather conditions (flanking, angle tie-in).

Fire Airbases—
Airfields of Aerial Firefighters

The entrance to the Aerial Fire and Smokejumper Base southeast of the town of Silver City, New Mexico. The facility is run by the U.S. Forest Service. During the annual forest fire season, there are at least two firefighting airplanes, one smokejumper crew, and one helitack crew stationed there.

It was hot on that day in the Gila Mountains, just a few kilometers north of the Mexican border. The sun shone mercilessly on the long-parched land. A warm, sometimes gusty wind blew steadily over the fields, woods, and wildlands. Base Manager Rance Irwin of the Silver City Airtanker and Smokejumper Base in southern New Mexico had already checked the humidity of the air in the morning, and it was clearly under ten percent! As sunny and beautiful as this day appeared to be, it was predestined for the outbreak of dangerous forest and surface fires.

At the airtanker base of the U.S. Forest Service (USFS), not far from the small town of Silver City, the crews of the firefighting planes (airtankers) stationed here, and the smokejumpers, had already prepared for a busy day. The heavy airplanes—one Type I DC-4 and one Type II SP-2H airtanker of the Aero Union—were long since ready for action, the reloaders had already filled the tanks with fire retardant. The men and women of the Silver City Smokejumpers, numbering some fifty persons, were just checking their equipment when the alarm bell rang loudly over the airport area. Each person here had unconsciously expected action very soon. Now it was reality—quickly but calmly the men and women went to the changing rooms to put on their special jump suits, strap on their parachutes, and take their backpacks with personal equipment.

The graphic shows the most important fire airbases in the ten western states of the USA and the four military bases of the MAFFS. The network of these firefighting airbases allows fast action by airtankers within 20 to 30 minutes.

A few minutes later, the jump plane that transported the smokejumpers had lifted off with fifteen jumpers and was flying toward Glenwood, a small town a few miles northwest of Silver City. Rance Irwin and two of his men sat in their small office and attentively listened to the speaker. Previously the observer at the Reno Lookout had reported a buildup of smoke to the north and called the USFS smokejumpers.

The big retardant tanks at the Aerial Firebase. Here up to 40,000 gallons (ca. 151,300 liters) of rusty brown extinguishing and cooling material are stored.

The personal equipment of the smokejumpers is kept in open cabinets in the changing room. The arrangement must be purposeful; nobody cares about exaggerated "living comfort."

A little later, the pilot of the jump plane confirmed the sighting. A fire had obviously broken out within the thick woods of the Apache Sitgreaves National Forest. The long distances and difficult access to the woodland area did not allow the forest firefighters to reach the scene quickly from distant localities. The leader of the smokejumpers thus urgently called airtankers from the Silver City Airtanker Base to prevent the fire from spreading if possible.

Naturally, the pilots of the two airtankers had expected this action. Soon the heavy engines started, the release to take off came over the radio, and the planes were on their way to the national forest.

This time, two retardant drops from the airtankers and the firefighting activity of the ground crews that arrived later were sufficient to get the forest fire under control. But the work of the Silver City Airtanker and Smokejumping Base crews was not finished yet—the hot, dry weather blanketing the landscape not far from the Mexican border resulted in two more firefighting calls for the men and women of the U.S. Fire Service.

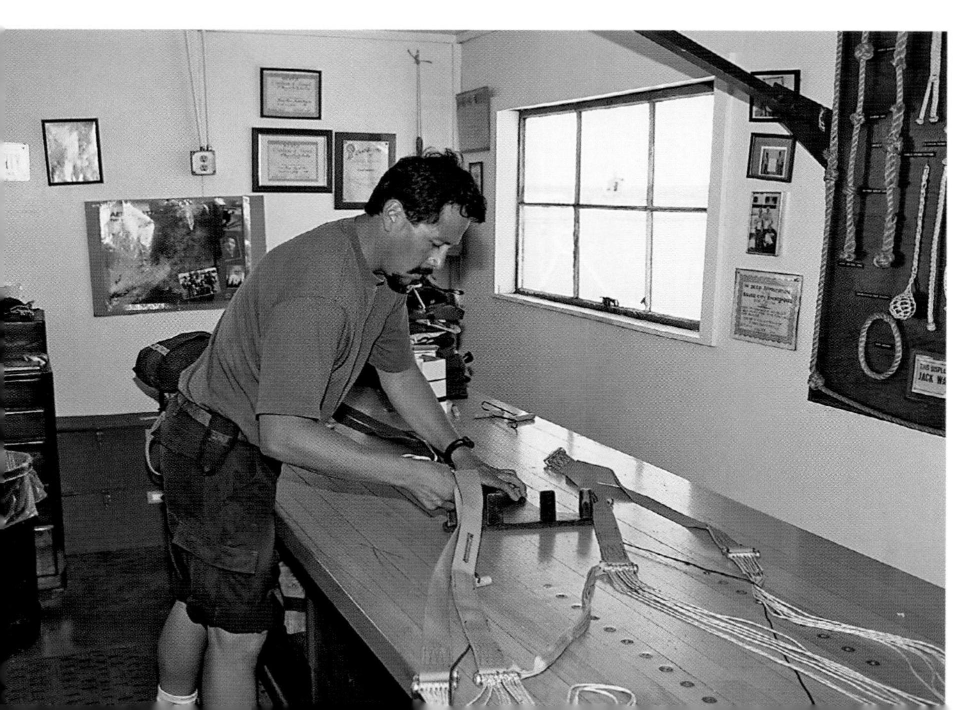

In their free time, the members of the smokejumping crew check their equipment and wait. There are about fifty smokejumpers at the Silver City Smokejumper Base during the annual forest fire season. This facility in New Mexico has been in existence since 1954.

Fire alarm! Calmly and without stress, the crewmen put on their jump suits.

U.S. FOREST SERVICE

Fully equipped, the smokejumpers from Silver City board the jump plane, a Douglas DC-3 owned by the USFS and numbered N115Z.

In the Aerial Fire Base office, Base Manager Rance Irwin (back) and two of his colleagues follow the radio traffic of the responsible command posts and the aircraft that are already in action.

This time the fire report came from Reno Lookout, a fire tower to the northwest, in Apache Sitgreaves National Forest. From there the smoke developing in the wooded area could be observed.

Two retardant drops from the airtankers are sufficient to stop the beginning forest fire this time. Ground crews of the U.S. Forest Service (USFS) took over the final work of extinguishing the fire.

Air Attack Bases (Fire Air Bases)

For the use of aircraft of all kinds in fighting forest and surface fires, airfields especially equipped for their uses are necessary. Their logistic facilities, the technical and personal setups, and the general aircraft and fire-fighting technological functions are specifically spelled out in the USA in the Interagency Airtanker Base Operations Guide (IABOG), which is obligatory for all national and state forest and fire protection organizations—along with regional requirements. A committee of the U.S. Forest Service (USFS, U.S. Department of Agriculture/USDA), the Bureau of Land Management (BLM, U.S. Department of the Interior/USDI), the Department of Natural Resources, and the California Department of Forestry and Fire Protection (CDF), in cooperation with the National Aviation Operations Officers (USFS, BLM) and the Airtanker Base Managers, meeting at an annual National Airtanker Base Workshop, turn out the volume of requirements, which numbers some 230 pages.

Other states that use aircraft for forest and surface firefighting have included the IABOG in their technical and tactical considerations, using the requirements as they are, or parts of them, or have modified them to fit their circumstances.

The Airtanker and Smokejumper Base of the National Interagency Fire Center (NIFC) in Boise, Idaho. NIFC is the central coordinating office for, among other things, firefighting that crosses state boundaries in the USA. From here, firefighting planes, personnel, and materials can be flown to all parts of the USA.

The Airtankers 99 (P2V) and 140 (P2V) at the Boise Airtanker Base in Idaho.

In practice, several kinds of Fire Air Bases (the general and common concept for Air Attack Bases) are defined:

- Airtanker bases for the use of firefighting airplanes (airtankers) that are loaded with fire retardant. Facilities of this kind are available mainly in the western USA and are maintained by federal and state forest and forest fire agencies.

- Airtanker bases for the use of firefighting airplanes (airtankers) that are loaded with water. Facilities of this kind are available in any countries of the world that use firefighting planes, and in the Mediterranean countries of Europe where amphibian firefighting planes are used. As a rule, they are maintained by national forest and forest fire agencies and by charter contractors.

- Heliports (or helibases) for the use of firefighting helicopters with flexible external load containers and helitankers with fixed tanks that are filled with water. Facilities of this kind exist internationally. As a rule, they are subordinate to the charter contractors, national forest or forest fire agencies, and regional or local (county or city) fire departments.

- Helispots, temporarily established helicopter bases which can also load helicopters with water. Helispots can be established by regionally responsible forest and forest fire agencies for either the duration of an entire forest fire season or especially for a single action. Helispots are usually located near open waters (rivers or lakes), so as to allow the filling of their tanks with Bambi buckets, or they offer the possibility of taking on water from containers (pumpkins) and/or firefighting vehicles (engines, water tenders) directly at the local base.

An Antonov 26 firefighting plane of the Russian Avialesookhrana at the central base in Pushkino near Moscow. Usually water is loaded at Russian airtanker bases, unlike American fire airbases, where fire retardant is usually used.

The firefighting plane of Avialesookhrana (A-26, #RA-26002) is being loaded with water. At the agency's own airbase, the machines are also serviced and loaded with fuel.

Helibases or heliports are stations for firefighting helicopters and helitankers. As opposed to the temporarily established helispots, they are permanent facilities that can be operated all year.
The Rifle Helibase at the edge of the Garfield County Regional Airport in Colorado, a small facility at which a firefighting helicopter is stationed, is run by the Bureau of Land Management (BLM) of the Upper Colorado River District.

A much larger facility is the Barton Helibase of the Air Operations department of the Los Angeles County Fire Department in Pacoima, outside Los Angeles. At the Barton Helibase, up to ten helicopters and helitankers are stationed, while workshops and servicing buildings complete the extensive facility, which even has its own airport fire department.

The so-called helispots are set up temporarily during the annual forest fire season. They can be established and assigned aircraft and personnel for the entire season or just for one action. This helispot is on the French island of Corsica in the Mediterranean Sea.

• Helitack bases as permanent stations of helitack crews (or rappelling crews) with their helicopters. At helitack bases there is usually no tank-filling apparatus, since the helitack copters are usually fitted with external Bambi buckets, which are filled just before going into action.

• Combined fire air bases, at which both airtankers and helicopters can be stationed—for example, the central base of the Securite Civile at the Marseilles-Marignane airport, where amphibian planes, airtankers, helicopters, and other aircraft for firefighting are stationed.

• Combined fire air bases, at which both firefighting aircraft (helicopters) and other aircraft for forest and surface firefighting (such as jump planes, lead planes, command planes) are stationed.

• Military (fire) air bases—in the USA, the 145th Airlift Wing (North Carolina Air National Guard), the 153rd Airlift Wing (U.S. Air Force, Cheyenne, Wyoming), the 302nd Airlift Wing (U.S. Air Force, Peterson, California), and the 377th Airlift Wing (U.S. Air Force, Kirtland, New Mexico)—at which the military (MAFFS) airtankers are stationed. All of these bases belong to the U.S. Air Force or the U.S. National Guard.

In other countries too, firefighting airplanes and/or helicopters of the military are stationed at military airfields.

A temporarily established helispot is used during a forest fire in Croatia. The helicopter, equipped with flexible external Bambi buckets (a Bell UH7 is shown), can be supplied with water here either from fire trucks or tank trucks, or from temporary water tanks.

A helispot is used during a forest fire in the Bohemian Switzerland National Park near Jetrichovice, Czech Republic. In the picture, two Czech police helicopters are using Bambi buckets, while three tank trucks (Tatra 815 and 813) supply the helicopters with water in the background.

A large helispot with several helicopters and helitankers in Arizona. Here no direct water supply is at hand, and the copters are usually supplied with water from nearby bodies of water.

A means of water supply in the Czech Republic: Bambi buckets are filled from a tank truck (water tender).

Three Czech Tatra 815 fire tank trucks with 8000-liter water tanks are ready to fill the helicopters' Bambi buckets at this helispot.

Helitack bases are locations of helitack or rappelling crews (here Helitack 555 of the Bureau of Land Management in Kern County, near Bakersfield, California). The 15- to 20-man crews each use a helicopter that flies them directly to the scene of action to fight the fire. There the crew can slide down ropes to the ground if the copters find no place to land. After the crew is landed, the copter is used for firefighting with Bambi buckets.

Helitack Crew 555 of the BLM from Kern County, California.

Setup and Equipment

Air attack bases are usually attached to public airports or military airfields. Thus they can use the technical and tactical infrastructure of the airport. All aircraft used for forest and surface firefighting (airtankers, helicopters, and other aircraft) are subject to the public air traffic regulations applying to the airport and the region, but usually have precedence over public air traffic when in action.

Air attack bases (fire air bases) are usually situated at the edge of an international airport but have direct non-public approaches (taxiways) to the public runways. Depending on the size of the air attack base, varying numbers of loading pits are set up there, each of them with its own usually loop-shaped approach way. Thus it is possible to load several aircraft with retardant simultaneously without their getting in each other's way or being delayed. As a rule, the taxiing and loading of an airtanker with retardant takes no more than eight minutes.

France's largest combined fire airbase is located at the Marignane airport near Marseilles (Aeroport Internationale Marseille-Provence). Along with the amphibian planes of the Securite Civile (Canadair CL-415), there are also various airtankers and helicopters stationed there. The picture shows the facility not far from the civil air terminal (upper right in picture), with the yellow CL-415 planes and the retardant filling apparatus (left). From the runway, the amphibian airtankers (scoopers) can go directly into the water of the Etang de Vaine to pick up water.

At the entrance to the Securite Civile air base is a mustered-out CL-215 (ex-Pelikan 44, #F-zf8V as an eye-catching monument.

Another view of the combined air base of Marignane, France, showing the parking area of the airtankers (S-2T Tracker) at left, the amphibian airtankers and larger airtankers (such as Dash 8, Fokker 27 and Convair 52 at right; the fire-retardant filling area is in the center.

The aerial fire airbase at Fresno, California.

- The standard prescribed response time (alarm to scene of action) in the airtanker contracts is 15 minutes after the completion of the loading process.
- The average time from the alarm to the beginning of the loading process is taken as three minutes.
- The average loading time for 3000 gallons of retardant (ca. 11,356 liters) six minutes (500 gallons per minute).
- Another two minutes are needed to remove the tank equipment and prepare the airtanker.
- This adds up to an average loading time of eleven minutes for Type I airtankers (3000 gallons).
- The total time from the alarm to action at the scene is thus some 26 minutes, depending on possible delays (for example, local air traffic, tactical planning time, technical checks).

For reasons of safety, airtankers should reach the scene of action only after a period of 30 minutes after sunrise (corresponding alarm) and should not be summoned 30 minutes before the official time of sunset.

The tactical center of an air attack base/ fire airbase in always the filling area with the large tanks of extinguishing and cooling materials, the mixing facilities, the supply and filling areas (pumps and hoses). Depending on the size of a fire air base, water, retardant, and chemicals are stored in varying quantities. The retardant tanks in which the ready-mixed cooling material is stored hold between 10,000 and 80,000 gallons (37,850 to 226,000 liters). At smaller fire airbases, mobile tank trailers can be used instead of immobile tanks.

From the retardant tanks, long filling hoses (2.5-inch diameter pipes, 400 gallons per minute) kept on spools are extended to the places where the airtankers stop. These hoses can thus be moved by one person, and the tanks of the airtankers can be filled by one person.

Fire retardant—a gel-like fluid to cool flammable materials—is mixed at the bases by various recipes of water with the addition of various chemical substances (including natrium, calcium, iron oxide) and suited to the individual action. The finished mixture is stored in the large tanks at the fire air base. In the tanks there is a stirring device that keeps the retardant ready for use, even over long periods of time.

The chemicals used to mix the fire retardant are delivered by the manufacturers in large plastic containers mounted on pallets, and stored near the tanks. Forklifts or built-in loading equipment move or lift the containers to the tank openings, through which the powdered chemicals are dumped into the water.

Other Facilities

Directions for the establishment of fire air bases (airtanker bases) are recorded—independently of the "Interagency Airtanker Base Operations Guide" (IABOG)—in the "Interagency Retardant Base Planning Guide" (IRBPG). The bases are rated according to their size in three categories for the action and storage buildings.

A look at the tower of the Fresno Fire Air Base shows the extensive area of the CDF-USFS facility at the Fresno City Airport. From here there are direct connections to the runways of civilian airports.

At the Fresno Fire Air Base at least two airtankers (at left, the Type III Airtanker 100 of the California Department of Forestry and Fire Protection/CDF, a CDF lead plane (center) and a command plane (right background)) of the U.S. Forest Service (USFS) are stationed.

The four retardant tanks at the Silver City Aerial Fire Base in New Mexico. The lettering shows that Phos-Check Type D 75 retardant is used here.

Retardant tanks at the Libby Airtanker Base at Fort Huachuca, New Mexico.Tanks of this size (standard retardant tanks) have a capacity of up to 10,000 gallons (ca. 37,850 liters).

This retardant mixing system is seen at the fire air base in Paso Robles, California—at left is the large mix container, at right the water container. By a hose connection (center), water can be added to the pulverized retardant material. The finished (gel-like liquid) retardant is pumped underground to the storage tanks and the filling apparatus.

The retardant supply setup (some 10,000 gallons/37,850 liters per tank) at the Minden-Gardnerville Fire Airbase south of Carson City, Nevada.

The filling and mixing tanks at the fire airbase in Grand Junction, Colorado. Here too, the Phos Check D 75 F retardant is generally used.

A smaller retardant mixing system at the Cedar City Air Tanker Base in Utah. The color-coded markings indicate water (blue) and retardant (red) intakes. The finished mixture is pumped into the storage tank.

A modern pump and dosier system is used for filling at the Fresno Airtanker Base in California. A modern Micro-Motion flow-through system by Emerson Process Management is installed to measure and regulate the flow of retardant to the filling rig.

The pump system at the air attack base in Paso Robles, California. A Super T Series centrifugal pump (3400 gallons/ca. 12.780 liters per minute, 215 liters per second) made by the Gorman-Rupp Company of Mansfield, Ohio, is shown at left.

The finished retardant is pumped into an airtanker's tanks through hoses (diameter pipes, 2.5 inch/6.3 cm) that are drawn out to the aircraft parking places. For better mobility, the filling hoses are mounted on rollers.

At some fire airbases, mobile retardant tanks are used instead of built-in tanks. Usually they are heavy semi-trailers that are left at the airbase during the forest fire season or are exchanged by the retardant delivery firms. At upper right, a mobile semi-trailer of retardant is seen at the Ramona Air Attack Base of the CDF in California. In the background is CDF Airtanker 71 #N432DF, a Grumman S-2T. In the picture below, a retardant-tank semi-trailer is shown at Cedar City Airtanker Base in Utah.

Hoses and couplings are standardized at all airbases and for all airtankers in the USA, so that any aircraft can be filled at any airbase. At left, the filler of the Airtanker 100, an S-2T of the CDF.

Below: A retardant tank hose coupling.

Size order for standard facilities (buildings, storage, equipment, filling sites) of air bases. (*) Reload fire airbases are occupied only during action in the region.

Size Order of Fire Air Bases		
Retardant Capacity (output)	**Action Buildings**	**Storage Buildings**
Reload (*)	1200-1500 sq.ft./110-140 sq.m.	400-800 sq.ft./35-75 sq.m.
Under 1,000,000 gallons	3000-3500 sq.ft./280-325 sq.m.	1000-1200 sq.ft./90-110 sq.m
Over 1,000,000 gallons	3500-4500 sq.ft./325-420 sq.m.	1500-2000 sq.ft./140-185 sq.m.

Every fire airbase has at least one office (sometimes also the radio room), storage rooms for equipment, materials, and retardant supplies, living space for crews (sometimes sleeping, sitting, and social rooms), a radio room, a group room for conferences, workshop rooms, and sometimes the corresponding space for a jump crew and-or a helitack crew.

In addition, the technical facilities of the fire airbase are also defined, including the radio, tele- and outside communication systems, the environmental protection facilities (especially the filling apparatus), and the safety systems.

Smaller fire airbases, such as this one at Cedar City, Utah, often have an "office hut" and their own living and sleeping trailers. The technical features are also "modest," but the capability of these airbases cannot be denied.

Fire Retardant

Several million gallons of fire retardant are loaded into airtankers and dumped on fire sites annually in the USA. Almost all American airtankers are loaded with retardant for firefighting. Loading with water or with water plus foam is done—to the extent that it is possible with the technical equipment—only in special cases, such as with amphibian airtankers.

European firefighting aircraft usually use water—including the Canadair CL-215 and CL-415 most commonly used there, as well as the Russian firefighting planes. Helicopters that are equipped with mobile external containers also use water, while the so-called helitankers (helicopters with fixed tanks) usually use water, but can also be loaded with retardant.

Fire retardant is usually not used tactically for extinguishing fires, but as a gel-like cooling material that effectively decreases the intensity of fires on the ground by covering them to reduce the air that reaches them.

Fire retardant (long-term retardant) as used in practice is a rusty brown gel-like fluid, mixed of, among other things, phosphates, sulfates, natrium-calcium borates and iron oxides (fire retardant chemicals of various manufacturers), and water.

The accommodations at the Libby Airtanker Base, on Fort Huachuca Military Base (U.S. Forest Service/USFS), are large and comfortable. There are various offices in the roomy building, plus a radio and communication center, living, sleeping, and activity rooms for the crew. Typical is the small "command balcony," from which the base manager can see the entire airfield.

Modest working, sleeping, and kitchen facilities in an old wooden building at the Cedar City Fire Air Base in Utah.

The radio center of the fire airbase in Fresno, California (CDF/USFS) is roomy and practical. The Dispatcher and action leader can work optimally from here.

The retardant components are delivered in big bags stored on pallets. Later their contents will be poured into storage or mixing tanks.

Various amounts of foam concentrates are part of a fire air base, along with the retardant. The picture shows a pallet of Class A Foam made by FireTrol (right).

Large trucks deliver the retardant containers to the airbases. The containers are then taken to the storehouse by the base's own forklifts or cranes.

The main ingredient of the retardant is a powdered "Phos Check."

Below: Fire retardant—a thick rust-brown gel-like liquid, composed of phosphates, sulfates, natrium-calcium borates (salts), and iron oxide dissolved in water. The finished liquid must constantly be stirred; otherwise it will settle as a thick "soup."

Fire retardant sticks not only to flammable materials of vegetation, but also, as shown here, to the human body. Any wildland firefighter who is hit by an airtanker drop of retardant certainly has a thorough job of cleaning himself to do.

Above: Workshops to service equipment, aircraft, and ground vehicles as well as their own facilities are generally part of the facilities of a fire airbase. This picture shows the small workshop at the base in Paso Robles, California, with its observation tower and radio communication.

In the large equipment building of the Silver City Aerial Fire Base, pallets of retardant and extinguishing material, and other materials, are stored.

Often so-called fire cashes are part of air bases. From these extensive stores of materials, many additional firefighters can be quickly supplied with the equipment, tools, and materials they need for action. Fire caches, depending on their size, can equip up to 5000 firefighters.

But one often sees this again and again too! Fuel and foam concentrate at a very small fire airbase (such as a reload base) may be stored at the edge of the facility, unprotected and yet problem-free.

Since August 2005, the California Department of Forestry and Fire Protection (CDF) has tested non-toxic Class A fire retardant gels with blue color additives, to make them more noticeable on the ground. Preparations are being made for further color tests.

There is a difference between short-term and long-term retardants. As a rule, long-term retardants are used on larger forest and surface fires, so that a longer cooling and covering effect by the material on the ground will create optimal conditions for the action of ground crews.

Long-term retardant can be mixed either by adding water to liquid fluid concentrate or by adding a powder product. All retardant mixtures are made to be as non-toxic and as harmless to the environment as possible, according to scientific testing and observation attested to by the authorities. Mixtures used on forest and surface fires can be suited to the conditions on the scene (vegetation, weather, topography, water) under these prerequisites. Special equipment (including refractometers, density meters, marsh funnels, and Brookfield viscometers) can measure and check the composition, nature, viscosity, and elasticity of the retardant mixed at the scene.

A look beyond the borders of the United States: European firefighting stations also use similar facilities. This is the aircraft workshop of the airbase in Eleusina, Greece. The picture shows a Grumman G 164 Ag-Cat.

This table gives an overview of the most often-used special types of fire retardant.

Fire Retardant (long-term)				
Name	**Type**	**Suitable for**	**1 ton of concentrate makes**	**Proportions: concentrate+ water = retardant**
Fire-Trol LCA-R	liquid concentrate	airtanker, heli-bucket	972 gal. (3679.42 liters) of usable retardant	1 gal. (3,79 Liter) + 5 gal. (18,92Liter) = 5,88 gal. (22,25Liter)
Fire-Trol LCG-R	liquid concentrate	airtanker, heli-bucket	923 gal.	1 gal. + 4.75 gal. = 5.67 gal.
Fire-Trol 931-R	liquid concentrate	airtanker, heli-bucket	962 gal.	1 gal. + 4.75 gal. = 5.71 gal.
Fire-Trol LCM-R	liquid concentrate	airtanker, heli-bucket	854 gal.	1 gal. + 4.25 gal. = 5.17 gal.
Fire-Trol LCP-R	liquid concentrate	airtanker, heli-bucket	853 gal.	1 gal. + 4.25 gal. = 5.15 gal.
Fire-Trol LCA-F	liquid concentrate	airtanker, heli-bucket	989 gal.	1 gal. + 5 gal. = 5.90 gal.
Fire-Trol LCG-F	liquid concentrate	airtanker, heli-bucket	933 gal.	1 gal. + 4.75 gal. = 5.66 gal.
Fire-Trol LCM-F	liquid concentrate	airtanker, heli-bucket	869 gal.	1 gal. + 4.25 gal. = 5.18 gal.
Fire-Trol LCP-F	liquid concentrate	airtanker, heli-bucket	864 gal.	1 gal. + 4.25 gal. = 5.14 gal.
Phos-Chek HV-R & HV-F (3.1:1)	liquid concentrate	airtanker	775 gal.	1 gal. + 3.1 gal. = 4.06 gal.
Phos-Chek HV-R & HV-F (3.6:1)	liquid concentrate	airtanker, heli-bucket	860 gal.	1 gal. + 3.6 gal. = 4.55 gal.
Phos-Chek LV-R & MV-R (3.6:1)	liquid concentrate	airtanker, heli-bucket	860 gal.	1 gal. + 3.6 gal. = 4.55 gal.
Phos-Chek MV-F (3.7:1)	liquid concentrate	airtanker, heli-bucket	881 gal.	1 gal. + 3.7 gal. = 4.64 gal.
Phos-Chek G75-F & G75-W	powder	ground, vehicle, heli-bucket	1907 gal.	1.12 pound + 1 gal. = 1.07 gal.
Phos-Chek 259-F	powder	airtanker, helitanker, heli-bucket, ground, vehicle	1869 gal.	1.14 gal. + 1 gal. = 1.06 gal.
Fire-Trol GTS-R	powder	airtanker	1325 gal.	1.66 gal. + 1 gal. = 1.10 gal.
Fire-Trol 300-F	powder	airtanker	1250 gal.	1.77 gal. + 1 gal. = 1.11 gal.
Phos-Chek D75-R & D75-F	powder	airtanker	1786 gal.	1.20 gal. + 1 gal. = 1.07 gal.
Phos-Chek WD 881 Class A Foam	additive	airtanker, helitanker, heli-bucket		
Fire-Sorb	structured viscose liquid	airtanker, helitanker, heli-bucket		
Phos-Chek Aqua-Gel (Europe: FocStop)	powder	airtanker, helitanker, heli-bucket, ground, vehicle		

A not exactly and properly mixed retardant can cause major problems and strongly influence the effectiveness of firefighting, for example:

- Reducing environmental protection through a too-high content of salt and/or poisons,
- Influencing the viscosity and resulting effect on the ground by increased crystallization,
- Causing abnormal flying conditions after the drop,
- Causing increased corrosion to tanks and aircraft.

If fire retardant, for example, contains too much water, it can result in:

- Decreased effectiveness on the ground (on the fire) through ineffective salt and chemical concentration,
- Drifting off during the drop,
- Separation of components and water during storage,
- Causing corrosion.

Checking the retardant quality thus is usually done at the beginning of the annual forest fire season, after the fire airbase is opened, after new deliveries of fire retardant chemicals, after finding abnormalities, and after the season ends at the fire airbase.

The use of fire retardant on forest and surface fires is generally not problem-free. It is agreed that retardant can have negative effects on the vegetation, ground water, and thus on the environment in general. Such problems are discussed at length in the USA. Scientific research institutes, under contract from the forest and forest fire agencies, plus the manufacturers of fire retardant, make every effort to decrease the environment-damaging qualities of their products through suitable combinations of components. This has largely succeeded—but there is no alternative to fire retardant.

Safety is another problem. Retardant drops from airtankers can endanger the ground crews. For example, directly hit equipment and vehicles can be damaged. People can be injured. Detailed procedural regulations and safety requirements for airtanker pilots and ground crews should assure that everyday routine measures like the dropping of fire retardant can be controlled and calculated by all involved parties.

Environment protection is emphasized at American fire airbases. Clear guidelines prescribe procedure with materials and dangerous substances. Many measures are as simple as they are effective. This homemade bucket guards well against remainders of retardant running out of the filling hose.

Here are some English terms related to the use of fire retardant and their definitions:

- **Component:** a part of the manufacturer's mixture of fire retardant.
- **Density:** the thickness of a substance per unit of mass.
- **Deterioration:** loss of viscosity within a certain time (such as 40% per year).
- **Dry concentrate:** a powder that is mixed with water to form the fire retardant.
- **Flow Conditioner:** a chemical added to the retardant in small quantities to prevent clotting.
- **Ingredient:** an individual chemical part of a manufacturer's fire retardant.
- **Inhibitor:** an ingredient of fire retardant to reduce certain chemical reactions (such as corrosion).
- **Liquid Concentrate:** unlike the powdered components of fire retardant, certain components are delivered in liquid form.
- **Long-term Retardant:** a series of retardant products that, with the addition of certain natural chemicals (salts), have a long-time effect. The salts work as retardants even after the water evaporates.
- **Lot:** a single delivery, usually 20 to 25 tons, of a retardant product or component, to a fire airbase.
- **Mixed Retardant:** ready-to-use fire retardant mixed with water and basic components.
- **Retardant:** an extinguishing or cooling material with chemical and/or physical effect to reduce flammability and the spreading speed of flames and heat.
- **Short-term Retardant:** a series of retardant products that cause their effects only in a wet or damp condition (including water). As soon as the water content of the retardant evaporates, the material loses its effect.
- **Specific Weight:** the weight in pounds of a gallon of fire retardant.
- **Thickener:** a chemical substance that increases the viscosity and/or elasticity of a liquid product. A component of fire retardant.
- **Viscosity:** the mass of a fluid condition of (partly) liquid materials, When fire retardant is manufactured, its viscosity is measured with a Brookfield Viscometer or Marsh funnel.
- **Viscosity Reducing Agent:** a substance (usually an enzyme) for reducing the fluidity of fire retardant.
- **Wet Concentrate:** a liquid concentrate used to make fire retardant, producing the finished retardant when water is added (see also "liquid concentrate").

Personnel

The crew of a fire airbase usually consists of the:

- **Base Manager**: Leader of the base and representative of the administrative organization, such as the BLM, and his deputy (Assistant Airtanker Base Manager), both responsible for the technical, tactical, and air-tactical operation of the base, for the safety of the operation (ground and air), and for the documentation of all actions.

- **Air Attack Officer:** the person responsible for the smooth operation of aircraft flights and their loading, unless – at smaller bases – this function is taken over by the base manager.
- **Retardant Mixing and Loading Crews:** Those who run the tank, mixing, and filling systems. Depending on the size of a fire air base, the mixing and loading of fire retardant may be tasks of the same crew.

Leaders who are directly subordinate to the base manager include the:

- **Ramp Manager:** The person responsible for the moving, loading, and parking areas of the airtankers and other action aircraft, as well as environmental protection in the loading area; subordinate to him are the Parking Tenders, who are responsible for all vehicle and personnel movements in the area of the parking and placement positions.
- **Retardant Mixmaster:** responsible for the preparation of retardant and for all technical equipment for producing and filling retardant, as well as its orderly loading into airtankers; subordinate to him are the Mixing and Loading Crews.

One of the base managers was Bill Parks (who died in April 2007), the chief of the Libby Airtanker Base in Fort Huachuca, New Mexico. A forest fire specialist appointed by the U.S. Forest Service (USFS), he served—like many other colleagues—for many years and left his impression on the concept of the fire airbase not far from the Mexican border.

The Retardant Mixing and Loading Crew at the Minden-Tahoe Airtanker Base in Nevada. If a certain hecticness arises in the course of a major action, it is frequently these crews who influence the smooth comings and goings of the airtankers through their concentrated and well-organized work and bring calmness to the entire operation.

The Air Attack Officer assures the smooth operation of aircraft arrivals and departures at a fire airbase—here, at the Fresno Fire Air Base in California, he talks with an airtanker crew. Before the planes take off, he (or in this case, she) gives the necessary instructions for the takeoff procedure.

The Retardant Mixmaster sees to the correct preparation of fire retardant by mixing water and chemicals. The right mix depends on, besides the firefighting and environmental conditions, the burning material (fuel), weather conditions, terrain features, and firefighting tactics.

Airtanker 60, a Type I Douglas DC-7, in loading position at a fire airbase (Redding, in northern California). While the loading crew takes care of loading the plane, the crew (in front at the bow wheel) makes a brief technical check. The air attack officer (right front) supervises the procedure and later releases the plane to the runway.

- **Aircraft Base Radio Operator:** responsible for the radio and telecommunications of the base and the aircraft.
- **Aircraft Timekeeper:** responsible for time management: flight times, loading times, servicing times, work times of the personnel, etc.
- **Retardant Contractor Manager:** working in cooperation with the manufacturers and deliverers of fire retardant, to maintain the supplies and arrange for deliveries.
- (Administrative) **Contracting Officer:** The (A/CO) is responsible for all contracts with aircraft charter firms (contractors). A/COs are not stationed at the fire airbases, but in regional offices of forest and forest fire agencies. A Contracting Officer (USDI, USDA) for all national airtanker contracts is at the National Interagency Fire Center (NIFC) in Boise, Idaho, while contracts for Alaska (USDI, OAS) are administered from Anchorage.

There are also—besides further specialists (such as training foremen)—service and support teams and administrative personnel, pilots and co-pilots stationed at all fire airbases, plus technical crews, servicing personnel, and the smokejumpers (up to 50) stationed at some bases.

Responsible for the qualification and the continuing training of the personnel according to all applicable requirements are the administrative organizations of the fire airbases (USFS, BLM, etc.). Initial and continuing training courses for all positions at a fire airbase are regularly offered and carried out by the aviation offices of the forest and forest fire agencies as well as the Office of Aircraft Service (OAS).

In Action!

In action, the Air Attack Officer gives the airtanker pilots instructions as to exactly when they are to take off or land. After landing, the airtanker goes to the loading pit and shuts off its motors or turbines there. The reloaders bring out the filling hoses and load the plane's tanks in, at most, eight to ten minutes. At the same time, the plane's crew examines the inside and outside of the plane and checks the vital technical functions. Only on instructions from the air attack officer does the plane restart its engines and leave the loading area on its way to the runway.

The Base Manager, as the leader of the facility and the deputy of the agency, is responsible for the smooth operation of all processes in the airbase realm (supervision, safety, documentation, supplying of retardant and fuel, sup-

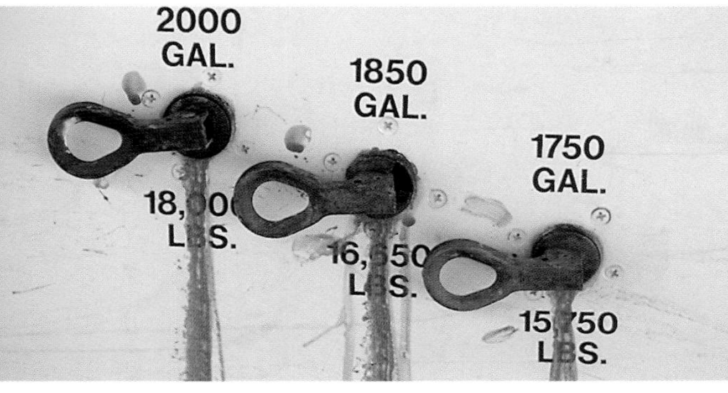

On Airtanker 16 (SP2H) the filling level of the retardant tanks can be read from the external indicators and overflow valves.

plying the personnel), and responsible also for the quick action of the firefighting planes and, when applicable, for the other aircraft (command planes, lead planes) and teams (helitack, smokejumpers). He and his colleagues also oversee all radio communications between the fire airbase, the tower of the public airbase, the aircraft in action, and the action command on site (air commander, incident commander).

In very serious situations and at suitable airbases, up to 15 airtankers can simultaneously land, be loaded, and take off. In technical terms, though, not all fire airbases can handle Types I and II airtankers.

The Hemet-Ryan Air Attack Base (USFS/CDF) in southern California is known as "the busiest tanker base in the world." During the annual forest fire season, there are at least four airtankers, two lead planes, and a helicopter stationed there. On average, about 1.1 million gallons of retardant (ca. 2.9 million liters) are loaded there during a season (the maximum was 3 million gallons in 1980!). The busiest day to date was in September 1985, when 13 airtankers flew a total of 61 flights within four and a half hours and dumped 70,400 gallons of retardant (ca. 266,500 liters). The maximum amount of retardant loaded in one day at the Hemet-Ryan Air Attack Base was 25,000 gallons (ca. 851,700 liters).

The reloader fills the airtanker's tanks with fire retardant. The loading of a large airtanker takes, at most, ten minutes. During or right after the loading, the loading crews also check the technical features of the plane, such as the filling levels, tank caps, tires, and wheels.

Type I Airtanker—The "Large Airtanker"

See also:
"The Future of Aerial Firefighting"—The End of the "Heavy Airtanker" in the USA

The American Incident Command System (ICS) divides airtankers into Categories I, II, III, and IV. These categories are based on the total capacities of the planes for extinguishing or cooling materials. Thus the Type I Airtanker (multi-engine airtanker/MEAT) is the largest type of firefighting airplane in the ICS classification. Outside the USA or outside the ICS system, planes of this size are called "large airtankers" or "heavy airtankers."

Type I airtankers—the best-known are the DC-7, the P3-A, and the C-130 MAFFS—are equipped with extinguishing- or cooling-material tanks with a capacity of at least 3000 gallons (about 11,355 liters). The KC-97 has tanks with a capacity of 4500 gallons (ca, 17m000 liters), while the Martin Mars, formerly used in Canada and now to be sold to Coulson Aircrane Ltd. holds over 7200 gallons (some 27,250 liters), and the Ilyushin 76 used in Russia holds over 11,000 gallons (some 41,635 liters).

See also:
More on "Large Airtankers" in "Firefighting Airplanes International"

At this time, there are four kinds of Type I airtankers operating in the USA: the Boeing KC-97 (Alaska), P3-A Orion of the Aero Union, the P2V (F5, 7) of Neptune Aviation and the Minden Air Corporation, and the C-130A MAFFS (military version), since most large airtankers (Types I and II) were retired for safety reasons by the American air oversight board (National Transportation and Safety Board/NTSB) and the U.S. Forest Service in May 2004.

Equipment

The large airtankers used in fighting forest and surface fires in the USA are generally rebuilt ex-freighters or military planes of the U.S. Air Force, which were taken over after being mustered out by the military.

This chapter portrays the best-known American Type I airtankers in their various functions—on the ground, in the air, while loading at a fire airbase, and in action over a forest fire. Most of the pictures are previously unpublished and portray the "big water bombers" in a unique and exciting manner.

Airtanker 60 (# N838D) is a Douglas DC-7 of the Aero Union Corporation charter contractor of Chico, California. The DC-7 is one of the few American firefighting planes that were not taken over from military service. These planes were originally used as airliners, and were then rebuilt as airtankers. The plane is 112 feet long (ca. 34 meters), and in comparison with the DC-4 and DC-6, which are also in use, is the largest version of the Douglas airtanker.

The DC-7 has been in service with the American forest and forest firefighting services since 1976. The DC-7 can be recognized by its square portholes, three of which are ahead of the wings.

Airtankers in action! Airtanker 20 (# N920AU) and Airtanker 26 (# N76AU), P3-A Orion Type I firefighting planes of the Aero-Union Corporation, are shown on the way to a fire and dropping their fire retardant over steep wildlands.

At right in the picture is another airtanker (AT-25, # N925AU) of the Aero-Union Corporation, at the Silver City Fire Airbase, along with Airtanker 06 (P2V-5, # N9855F) of the Neptune Corporation, from Missoula, Montana.

A P3-A Orion of the Aero Union Corporation, Type I Airtanker 26, pictured at the Redding Fire and Smokejumper Base in California. The four-engine turboprop plane crashed on a training flight in the Brushy Mountain Area in California in April 2005. The three crew members were killed in the crash.

26

It ranks among the largest US firefighting planes—Airtanker 63, a Lockheed C-130 Hercules of the TBM Corporation. The C-130 was formerly a military freight and transport plane, which after its conversion has been used for forest and surface firefighting in two versions: as a conventional Type I airtanker with a capacity of 3000 gallons (ca. 11,355 liters), and fitted with the Modular Airborne Fire Fighting System used in military planes of the U.S. Air Force and U.S. National Guard.

The crew of Airtanker 63: from left to right, co-pilot Andy Kremer, pilot/captain Arch McKinley, and flight engineer Patrick Kucera.

The picture at the left shows the interior of the gigantic freight area with the 3000-gallon (11,355-liter) tank. The standpipe that shows externally how full the tank is can be seen clearly.

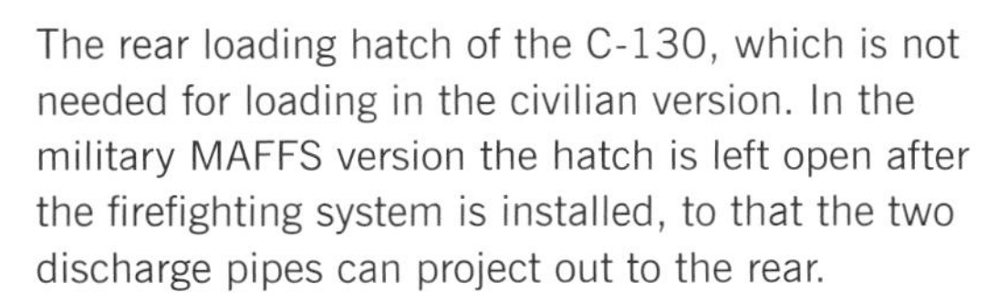

The rear loading hatch of the C-130, which is not needed for loading in the civilian version. In the military MAFFS version the hatch is left open after the firefighting system is installed, to that the two discharge pipes can project out to the rear.

64
N466TM

64

Airtanker 64 (# N466TM) is another large firefighting plane (large airtanker) of TBM Inc., at the Fresno Fire Airbase in California.

Lower left: Airtanker 64 taking off from the airport in Fresno, California.

A third airtanker of this type, Airtanker 67 of the Central Air Service (# N96541), dropping fire retardant over a heavily forested area.

Airtankers 130 (# N130HP), 131 (# N131HP), 132 (# N132HP), 133 (# N133HP), and 134 (# N134HP) were prepared by Hawkins & Powers Aviation in Greybull, Wyoming. These are Lockheed C-130A planes that are classified as Type I airtankers with a load capacity of 3000 gallons (ca. 11,355 liters). Unlike the gray paint with yellow number stripes, Hawkins & Powers paints its firefighting planes cream with wide red stripes.

Airtanker 130 is shown in July 2001 at Libby Airtanker Base in Fort Huachuca, New Mexico. This plane crashed barely a year later, on June 15, 2002, while fighting a fire near Walker, California, and its crew lost their lives. Its action at Walker and the circumstances of the crash are described in detail in "The Future of Aerial Firefighting—T-130 down!"

The gigantic rear hatch of the C-130, which the military originally used for loading transport machines, is imposing. When it was lowered, vehicles could be driven up the ramp and into the plane.

Measuring 133 feet (ca. 40.50 m), the wide wing of the C-130 and the four turbine propeller (Allison T56-A-9D) engines get the aircraft flying up to a maximum airspeed of 275 mph (ca. 442.6 km/h).

The cockpit of the C-130 is also huge, and the front windows offer optimal views to the sides.

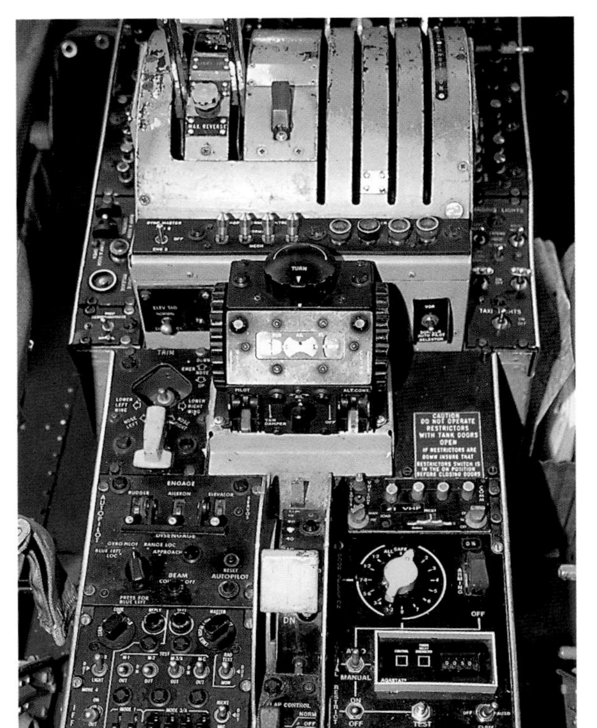

The central console of the plane, which was put into service in 1957. From here the pilots controlled the turbines (right front) and many other functions of the airtanker.

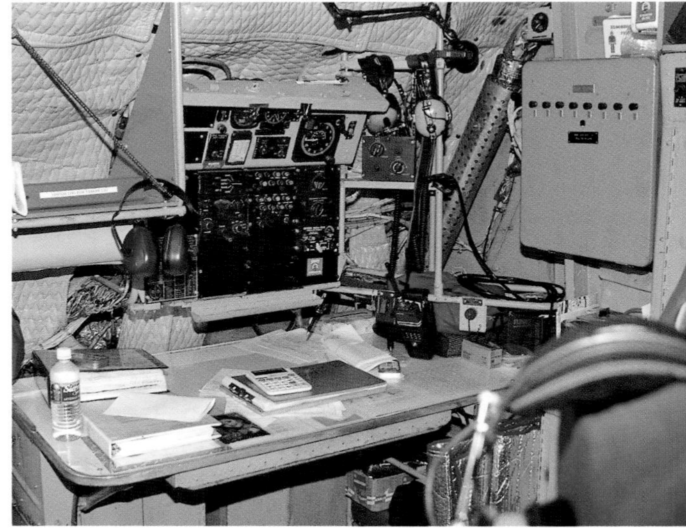

Behind the pilot's seat or over the lower floor of the plane is the navigator's workplace.

On the lower floor—the former freight area—the tanks are mounted and secured from sliding by strong chains. The entire electro-technical and hydraulic installations are mainly open.

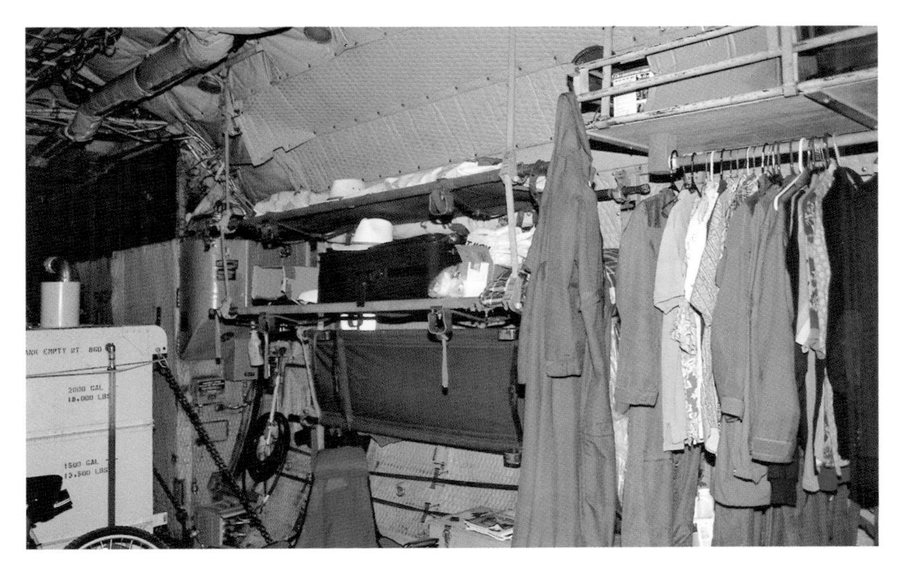

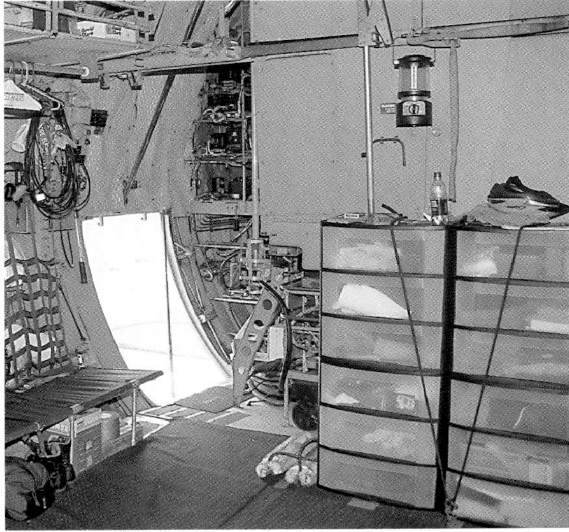

The loading space of Airtanker 130 (# N130HP) also serves as the crew's temporary living, changing, and working room during the forest fire season. Here personal effects are stored, plus food, clothing, and sports gear. Thus the crewmen are independent of the accommodations at the individual fire airbases, and the season lasts several months in the USA, during which the crewmen are on their own.

Another Lockheed C-130A, Airtanker 131, of the Hawkins & Powers Aviation firm, is seen dropping fire retardant.

Type II Airtankers—The Small "Large Airtankers"

Type II airtankers also rank among the "large or heavy airtankers, but form a lower category in the Incident Command System with their tank capacity of 1000 to 2999 gallons (ca. 6800 to 11,300 liters).

In American aerial firefighting, the Type II airtankers represent the largest numbers of firefighting planes. Among the best-known tankers in this category are the Douglas DC-6, Douglas DC-4 (C 54), Lockheed SP2H, and Lockheed P2V Neptune.

The firefighting planes of this size were also mainly taken over from the military and converted, rebuilt or extended for use in fighting forest and surface fires. One of the few planes also taken over from civilian airlines is the DC-4, which is very often used as an airtanker in the USA.

The Type II airtankers were also restricted from forest and surface firefighting as of April 2004 for reasons of safety. At this time only a few planes of this category are in use. They had to undergo an especially intensive and detailed safety test in advance.

Generally long since mustered out are the C-119, a twin-engine military transport plane of the U.S. Air Force, which was nicknamed the "Flying Boxcar" because of its great load capacity, the C-119's successor, the Fairchild C-123, also a transport plane used by the U.S.

Air Force and Coast Guard, the Consolidated PB4Y2 Privateer (later upgraded as Super PB4Y2), a long-range bomber of the U.S. Navy, which had been used for many years as an airtanker, the Boeing B 17 Flying Fortress, an ex-World War II Air Force bomber, and the North American AJ (A-2 Savage), a three-seat bomber of the U.S. Air Force.

These planes were also very well known in the postwar years and into the 1970s, and were often used for firefighting. They characterized the image of American aerial firefighting.

The Douglas DC-4 Skymaster (military designation C-54 E) is the smallest airtanker of the Douglas DC-6 and DC-7 types. These planes were at first used in the civilian realm as passenger and freight planes, as well as in the military. The DC-4 is slower than most other airtankers, but can operate from smaller and weight-limited fire airbases. The DC-4s, with its rounded cabin windows (those of the other DCs are rectangular), are very popular among flying firefighters in the USA, and are widely used as firefighting planes. In this picture, Airtanker 160 of Aero Flite, Inc., from Kingman, Arizona, is seen at the Reno-Sparks Fire Air Base in Nevada. Airtanker 160 can be loaded with 2000 gallons (7.570 liters) of retardant, which can be released through 4 to 8 hatches. The wingspan is 118 feet (ca. 36 meters); the top speed is some 220 mph (ca. 354 kph).

Airtanker 160, a DC-4 with number N96358.

Airtanker 119 (C-54G, # N406WA) of ARDCO Inc. of Tucson, Arizona, is also a Douglas DC-4.

Also run by ARDCO Inc. is Airtanker 151 (C-54E), a very plain-looking DC-4 with number N460WA. Douglas DC-6 airtankers have four engines with three-bladed propellers.

Airtanker 161 (DC-4/C-54E), a Type II airtanker, (# N82FA) of Aero Flite Inc. of Kingman, Arizona, is seen after landing at Redding Fire Air Base in California.

Airtanker 12 (# N96264), a P2V-7 of Neptune Aviation Services, Inc. of Missoula, Montana, is seen dropping retardant.

Right page: Also among the American Type II airtankers is the Lockheed SP2H. It is a structurally modified Lockheed P2V Neptune. The SP2H has been used to fight forest and surface fires in the USA since 1987. In the rebuilding, the planes were fitted with turboprop engines in place of the original piston engines, plus a retardant tank and system with a capacity of 2000 gallons (7570 liters). The system is computer-controlled.

The wingspan of the SP2H measures 98 feet (ca. 30 meters), and its top speed is 220 mph (ca. 354 kph).

In this picture, Airtanker 01 (# N701AU) is seen at the fire airbase in Porterville, California. The former U.S. Navy sea reconnaissance plane is run with striking red-white-black paint by the Aero Union Corporation of Chico, California. Below: Airtanker 01 in flight.

A Douglas DC-6 (Airtanker 68, # N90739), of TBM Inc., is seen in service near Santa Fe, New Mexico. The DC-6 is the medium version of the Douglas firefighting planes used in the USA. The plane has a wingspan of 118 feet (ca. 36 meters), and its top speed is 240 mph (ca. 386.2 kph). Its retardant tanks hold 2400 gallons (ca. 9080 liters).

01
01

01
N701AU
01

Also run by the Aero Union Corp. is Airtanker 16 (SP2H, # N716AU), seen here at the air attack base in Paso Robles, California. The Aero Union Corp. also runs the identical Airtankers 01 (# N701AU) and 18 (# N718AU).

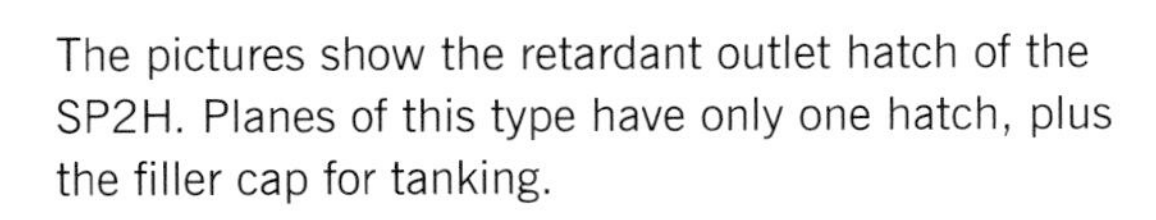

The pictures show the retardant outlet hatch of the SP2H. Planes of this type have only one hatch, plus the filler cap for tanking.

Airtanker 12 (P2V-7, # N96264) of Neptune Aviation Services. The U.S. Forest Service was the first government forestry and forest fire agency that took over P2V planes from the U.S. Navy and fitted them with six-hatch tanks for dropping fire retardant. Later the planes with the two jet powerplants proved to be very capable in their tactical tasks of fighting forest and surface fires, especially in rugged country with poor visibility. The planes are available in P2V-5 and P2V-7 versions.

Airtanker 11 (# N14447), a P2V (P2-H) of Neptune Aviation Services, Inc. This firm from Missoula, Montana, runs this plane plus the identical Airtanker 05 (P2V-5, # N96278), 06 (P2V-5, # N9855F), 07 (P2V-5, # N807NA), 08 (P2V-7, # N14835), 09 (P2V-7, # N4235T), 10 (P2V-7, # N4235N), 11 (P2V-7, # N14447), and 12 (P2V07, # N96264). The P2V planes were patrol craft and submarine chasers of the U.S. Navy until the mid-sixties.

Airtanker 06 (P2V-5, # N9855F), of the same charter contractor, is seen at the loading pit of Silver City Fire Air Base in New Mexico.

Airtanker 06 is flown by pilot Tom Raider and co-pilot Kris McAleer.

Airtanker 05 (P2V-5, # N96278) of Neptune Aviation Services, Inc. is seen at the fire airbase in Porterville, California.

RESCUE
NEPTUNE AVIATION SERVICES, INC.
05
MISSOULA, MT

08

At the fire airbase in Grand Junction, New Mexico, Airtanker 08 (P2V-7, # N14835) of Neptune Aviation Services, Inc. is parked. The firm has a branch in Alamogordo, New Mexico.

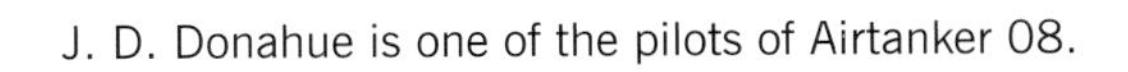

J. D. Donahue is one of the pilots of Airtanker 08.

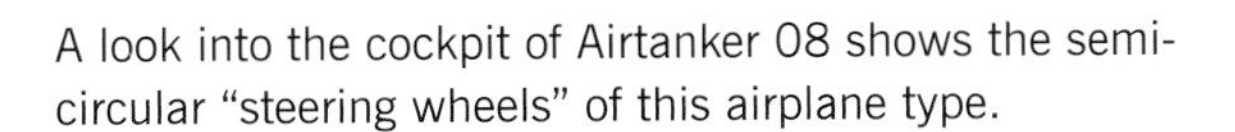

A look into the cockpit of Airtanker 08 shows the semicircular "steering wheels" of this airplane type.

"Ready for takeoff!" for Airtanker 99 at Cedar City Airtanker Base in Utah.

T-99 (# N299MA) is a P2V-7 of the Minden Air Corp. of Nevada. In October 2003 the plane crashed in the East Highlands of California during a return flight from a fire in Arizona. Both crewmen were killed.

Taking off in Airtanker 99 in the co-pilot's seat—surely a special experience. This made the plane's later crash particularly sad.

The retardant tank has been built into the freight area of the P2V-7. It holds 2400 gallons (ca. 9080 liters) of red-brown extinguishing and cooling fluid.

A look under the plane shows the tank outlet.

Airtanker 140, also a P2V-7, is run by the Hawkins & Powers Aviation firm. The pictures show the plane at the Boise Airtanker Base in Idaho. The charter firm from Greybull, Wyoming, also runs the same-type Airtanker 139 (# N139HP).

The consolidated PB4Y2 Privateer is one of the most striking firefighting planes in American aerial firefighting. The World War II heavy bomber, built in the 1940s, was used by both the U.S. Army (as B-24) and the U.S. Navy (as "Privateer"). In the mid-sixties, several privateers were taken over for fighting forest and surface fires. During the rebuilding, the powerplants were replaced by more powerful engines—the new version is called PB4Y2.

The PB4Y2 Privateer was rated as a Type II airtanker with a load capacity of 2200 gallons (ca. 8326 liters). The retardant tank had 6 to 8 dumping hatches. The wingspan measured 110 feet (ca. 33.50 meters), the top speed was 210 mph (ca. 338 kph). These planes have now been mustered out, as the Hawkins & Powers Aviation charter contractor finally ran out of spare parts for servicing their five identical airtankers: AT-121 (# N2871G), AT-123 (# N7620C), AT-124 (# N2872G), AT-126 (# N7962C), and AT-127 (# N6884C).

Airtanker 121 (# N2871G) is seen beside a lead plane of the charter company.

Airtanker 123 (# N7620C) at the Hawkins & Powers airfield in Greybull, Wyoming.

Airtanker 127 (# N6884C) in flight over Wyoming.

Type III Airtanker—The "Medium Airtanker"

Type III airtankers are the medium-size firefighting planes (medium or standard airtankers) in the American Incident Command System (ICS). The planes carry retardant tanks with a capacity of 600 to 1799 gallons (ca. 2270 to 6800 liters).

While the Type II planes were the most often-used standard airtankers in the American West until the retirement of the large airtankers in April 2004, this designation has become much clearer since that controversial decision by the U.S. Forest Service (USFS/USDA). The "medium" firefighting planes had to replace more and more of the large airtankers. In California in particular, the Type III airtankers of Type S-2T (formerly S-2), along with helicopters, have taken over the greater part of forest and surface firefighting from the air.

Airtanker 76 (# N417DF), an S2A made by Grumman Aerospace of Bethpage, New York, was a submarine hunter in use by the U.S. Navy from 1952 to 1972. In 1972 the CDF took over 19 S-2As from the American Defense Department and used them since 1973 for fighting forest and surface fires in California. Since then, these small and nimble airtankers with their 800-gallon (ca. 2028-liter) retardant tanks are an efficient part of the agency's forest fire units. The S-2As were equipped with two nine-cylinder propeller-turbine engines of the Wright R-1820-82 type, which produce 1500 HP each. Their top speed is some 195 mph (ca. 314 kph).
The S-2A was replaced as of 2003 by the first fifteen more modern and powerful S-2T Trackers. The Type III airtankers are used mainly for initial attacks on forest and surface fires.

Among the best known firefighting planes of this class are also the American Douglas A-26 Invader (later B-26), an ex-U.S. Army twin-engine bomber, the Fokker F-27, used in Canada as well as in special cases in the USA, and the AT-802 and AT-802 Fire Boss, further developments of former agricultural planes that are taking on an increasingly significant tactical role in forest and surface firefighting in the USA as well as in Europe and many non-European countries. Also ranked among the Type III airtankers are most of the amphibian firefighting planes (except the Russian Beriyev and the Canadian Mars, plus the Canadian Canadair CL-215 and CL-415 Bombardier airtankers used mainly in the USA, Canada, and Europe, and the American Consolidated PBY Catalina and Canso.

A Type III airtanker of the S-2T type of the California Department of Forestry and Fire Protection (CDF) is seen dropping fire retardant.

Airtanker 77, an S-2A with number N423DF, is dropping retardant.

A formation flight of CDF Airtanker 72 (# N435DF) over the Pacific coast. The California Department of Forestry and Fire Protection presently uses 23 Type III planes of the Grumman S-2T Tracker type, numbered 70 through 83, 86 through 96, and 100. Some of the numbers duplicate those of the federal government's airtankers.

An S-2A of the Redding Air Attack is being filled with fire retardant.

Redding Air Attack
78
HOT LOAD
N41

Airtanker 78 (S-2A, # N412DF) is stationed at the Redding Air Attack Base.

Airtanker 100 (S-2T Tracker, # N436DF) is one of the modern firefighting planes of the CDF. In 1996, 26 of these planes were taken over from the U.S. Department of Defense and converted to firefighting planes with 1200-gallon (ca. 4540-liter) retardant tanks. AT-100 is stationed at the Fresno Airtanker Base of the CDF/USFS.

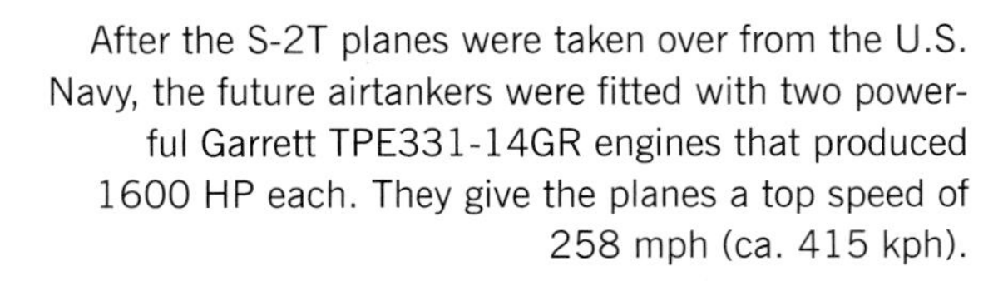

After the S-2T planes were taken over from the U.S. Navy, the future airtankers were fitted with two powerful Garrett TPE331-14GR engines that produced 1600 HP each. They give the planes a top speed of 258 mph (ca. 415 kph).

Vito Orlandella, one of the pilots at the Fresno Airtanker Base, in the cockpit of his S-2T Tracker. Right behind the cockpit are drawers and compartments for the onboard technical equipment, and the retardant tank is in back.

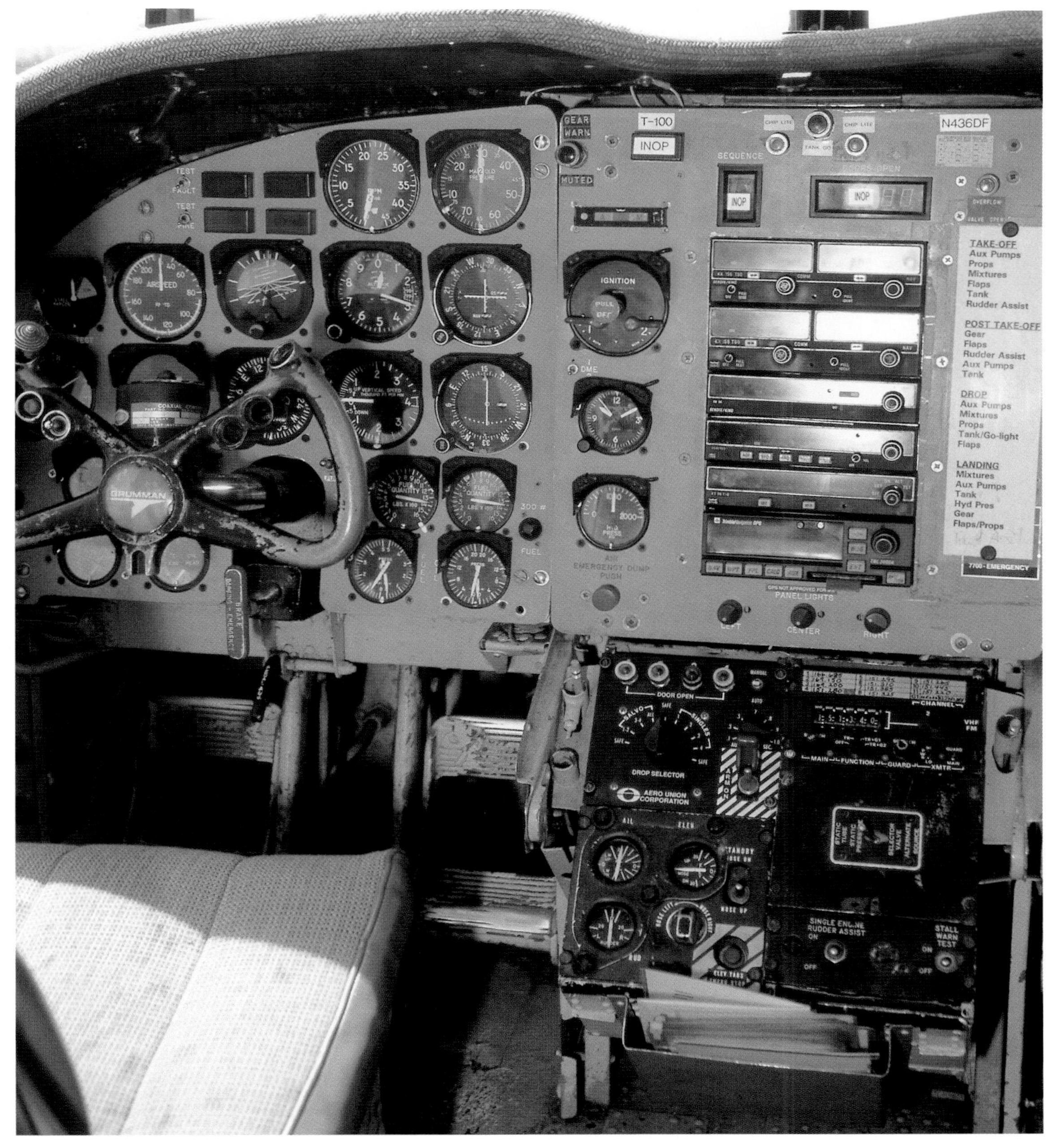

The cockpit of the S-2T Tracker, with the well-organized instruments of Airtanker 100. In action, the plane is flown and the retardant dropped by the pilot alone.

Among the Type III airtankers are the modern AT 802 and AT 802 Fire Boss planes (at the left front is the Canadian Conair Airtanker 82) of the Air Tractor firm of Texas. Photos of other planes of this type and the Type IV airtanker by Air Tractor can be found in the Single Engine Air Tanker (SEAT) chapter.

Type IV Airtanker—The "Small Airtanker"

The Type IV airtankers are the smallest firefighting airplanes (small airtankers), after all those planes of classes I to III already described (large and medium airtankers), that are used in the USA to fight forest and surface fires, or were in the past. According to the American Incident Command System (ICS), these are planes having tanks with capacities of 100 to 599 gallons (ca. 380 to 2270 liters).

Among the best-known aircraft in this category are the formerly agricultural planes (Ag-airtankers) that were rebuilt to fight forest and surface fires, including the PZL Dromader developed in Poland, the Thrush (previous model) or Ag-Cat.

A further development of the modern Ag-airtanker is seen in the airplanes built by the Air Tractor (AT) firm, Types 400 (AT-401B, AT-402A, AT-402B), 500 (AT-502A, AT-502B), and 600 (AT-602). Also ranked among the Type IV airtankers are the DHC2 Beaver, used in Canada and occasionally in the northeastern United States, a small but very robust airplane that was also a successful firefighting plane in amphibian form.

At this time, Type IV airtankers see service in the USA only in rare cases and in initial actions against small surface fires. These planes are used much more often in Canada (DHC2 Beaver), in the Mediterranean countries (including Spain, Greece and Croatia), and in numerous East European countries (including, among others, Poland, the Czech Republic, and Slovakia). The number of Air Tractors used in forest and surface firefighting there has increased greatly in recent years.

Agricultural airplanes were used in the USA, Europe, and other countries after World War II for fertilizing and spraying insect pests from the air. After about the 1950s, the Ag-tankers were developed into firefighting planes for use against small vegetation fires.

An Ayres S2R-T45 Turbo Thrush in action over Safford, Arizona. The SEAT plane is run as Airtanker 414 (# N3299H) by Aircraft and Equipment holdings of Phoenix, Arizona.

A so-called Single Engine Air Tanker (SEAT, Type IV airtanker) in flight over Big Bear Canyon in Florida. These smallest of all airplanes used to fight forest and surface fires were developed from former agricultural planes (Ag-tankers). In this picture, is Airtanker 493 (AG Pilots, Inc., # N9046D), an Air Tractor AT-602, which crashed in Clarksville, Florida, in December 2000.

Quite rare in the USA, former agricultural airplanes are often used in Eastern Europe. The pictures show a Czech Zlin 137T Agro Turbo (# OK-EJA) made by the Czech firm of CzechAircraft s.r.o.—Zlin Aircraft, in Otrokovice.

Often used in southern and eastern Europe, and at times in the eastern states of the USA, are the small airtankers of the PZL-Mielec M-18 Dromedar type, made by the Rockwell International consortium, and likewise former agricultural planes (Ag-tankers). The picture above shows a 1982 Dromedar from Greece, with a tank capacity of 581 gallons (200 liters) of water. Below is a PZL Dromader M-18A of Avialsa from Sagunto, Spain.

A New Generation—The "Single Engine Air Tanker" (SEAT)

See also:
The Future of Aerial Firefighting—Conversion—Developments and rebuildings of aircraft for forest and surface firefighting.

After the American "large airtankers" (Types I and II) were retired in May 2004, the leadership of the national and state forest and forest fire agencies faced the question of what aircraft could effectively replace the large airtankers. After long and controversial discussions, the two national agencies: the U.S. Forest Service (USFS/USDA) and the Bureau of Land Management (BLM/USDI), plus the DOI agencies (BLM, NPS, BIA, OAS) of the Department of the Interior (USDI), in cooperation and agreement with the U.S. Department of the Interior, decided in favor of small single-engine, single-wing airplanes, the so-called Single Engine Air Tanker (SEAT).

Airplanes of this kind have been developed from agricultural planes since the end of World War II, and had led a sort of second-class existence in the area of forest and surface firefighting in the USA up to this time. After the sudden, fast, and, in the technical world, unexpected decision of the U.S. Forest Service (USFS) on the retirement of large airtankers, action had to be taken, and so the U.S. Department of the Interior set up a SEAT program that suggested a large number of SEAT planes be sent to the American states and regions as replacements for the eliminated Type I and II airtankers. Although the SEAT program was objected to, primarily for tactical reasons, by many regional fire managers, the number of Single Engine Air Tankers used in forest and surface firefighting increased considerably. Nearly every state stationed these small planes, sometimes a number of them, at the first "abandoned" fire airbases, so as to combine them with the still-present helicopters and helitankers in a new and changed tactical concept of aerial firefighting.

After a number of the large airtankers had meanwhile been put back into service in the USA, the whole concept of forest and surface firefighting from the air (aerial firefighting) with Types I to IV airtankers, SEATs (some Type III, some not ICS-categorized), and firefighting helicopters and helitankers (Types I to III), provided a thoroughly functional and significant potential for supporting surface firefighters.

There were also conceptual considerations from the Department of the Interior (DOI) and the charter contractors for the purpose of forming a long-term aviation management entity. This particularly included a series of tests with firefighting aircraft/airtankers that could carry and drop more of the hitherto common extinguishing and cooling materials (such as the Boeing 747 of the Evergreen firm).

In non-American countries, the use of single-engine airtankers was much more widespread. This was true, among others, of the southern European countries and in many other parts of the world (for example, Africa, South America, and Australia). East European aircraft builders in particular developed—likewise from agricultural aircraft already in use for a long time—powerful firefighting variants with sometimes considerable potential for carrying extinguishing materials. Leading the way in Eastern Europe were the Polish firm of Panstvove Zaklady Lotnicze of Warsaw and several branches of it, such as PZL-Mielec, PZL-Warszava-Okecie, PZL-Svidnik, and PZL-Bielsko. In the USA, the Air Tractor manufacturing firm of Olney, Texas, was able to score particular success with new, modern products that found much use in national and state SEAT programs.

Air Tractor—An Example of Development

A particular example of the development of agricultural airplanes into modern airtanker production is provided by the Air Tractor Inc. firm of Olney, Texas. Their founder, Leland Snow, got involved in the technical developments of the postwar years and produced agricultural airplanes in the 1950s.

As of 1951 he designed and developed the Snow-1 (S-1), in the 1960s the S-2R Thrush, and as of 1972 the first Air Tractor (AT-300). Numerous other types and versions followed later. All of these planes, especially the Air Tractors, were not only very successful in the growing field of agricultural flying at the time, but also, and to this day, in aerial firefighting.

The present-day, modern Single Engine Air Tankers (SEAT)—AT-802 and AT-802 Fireboss—were meanwhile put into service in the USA, Europe, and many other lands as Type III airtankers for the fighting of forest and surface fires. Particularly after the retirement of the large Type I and II airtankers in the USA, Air Tractor was able to secure a firm place in international aerial firefighting. Today Air Tractor produces numerous agricultural and firefighting airplanes of Types AT-402, AT-502, AT-602, and AT-800, with tank capacities indicated by the type numbers. In 1996, the production facilities in Texas were expanded by 55,000 square feet (some 5100 square meters), and the 1400th Air Tractor was built. In the next year the number of delivered planes rose to more than 2000, and in 1998 Air Tractor celebrated the 40th anniversary of the Olney, Texas, firm.

Air Tractor's AT-802F Fire Boss dropping water over a fire.

A European representative of Air Tractor, Air Tractor Europe, is located in Sagunto, near Valencia, Spain. Among the firm's numerous customers are agricultural, forestry, and firefighting agencies in the USA, Canada, Mexico, and South America as well as South Africa, Australia, New Zealand, Spain, Croatia, Saudi Arabia, and Korea. The present chapter on the Single Engine Air Tanker (SEAT), plus the chapter on aircraft for forest firefighting in Europe will also offer much information on the small airtanker.

"Powerful action!"—An Air Tractor AT-802F on a computer-guided water drop. The tank of the Single Engine Air Tanker (SEAT) holds 820 gallons (ca. 3100 liters), which places the AT-802F among the Type III airtankers in the USA. It also has an 18-gallon (ca. 68 liters) foam tank. The small, modern planes are reserved, not only in the USA, for initial action in forest and surface fires and in fighting so-called spot fires on the edges of larger fires.

The 800 series is fitted regularly with an 800-gallon water tank. The wingspan of the plane measures 58 feet (17.67 meters). The basic version is driven by a PT6A-65AG motor.

Single Engine Air Tankers (SEAT) are being used more and more in European countries as well. In this picture at the Greek military airport of Tatoi we see a PZL M-18 Dromader made by the Polish firm of Polskie Zaklady Lotnicze (PZL-Mielec), which was developed and built according to Western standards in conjunction with the US consortium of Rockwell International. The M-18 has a tank capacity of up to 2500 liters (660 gallons).

A Spanish Single Engine Air Tanker (SEAT) made by Air Tractor.

A SEAT 103 is being serviced and repaired at the Tatoi military airport in Greece.

The Czech Single Engine Air Tanker (SEAT) with OK-EJA registration is seen at the Otrokovice airport west of Zlin. It is a ZLIN Z 137T Agro Turbo, made by Morovan-Aeroplanes (License CzechAircraft s.r.o./ZLIN, Otrocovice/Czech Republic).

Airtanker 1269, a Grumman G 164 with an 1890-liter water tank.

ZLIN Z 37a Cmelak (registration OK-AGR), a Czech Single Engine Air Tanker (SEAT) of the Agro Air firm. Planes of this kind were originally conceived as agricultural planes before they were adapted for firefighting tasks. The "Cmelak" (Bumblebee) was also built under license by ZLIN Aviation. The plane is 8.60 meters long, has a wingspan of 12.20 meters, carries a 700-liter water tank, and reaches a top speed of 200 kph.

ZLIN Z 137T Agro Turbine SEAT in flight.

The AT-802F is presently the most modern and powerful Single Engine Air Tanker (SEAT) made by the American Air Tractor firm. The Fire Boss differs from the standard 802 version in having two pontoons, making it a light amphibian plane. The plane is fitted with an 800-gallon (ca. 3028-liter) tank for extinguishing fluid.

EC-IVM

Fire Boss

The AT-802F Fire Boss made by Air Tractor is a modern bold but effective design of a Type III (medium) airtanker. As an amphibian airplane, the Single Engine Air Tanker (SEAT) can take off from and land on bodies of water.

The Modular Airborne Fire Fighting System (MAFFS)

MAFFS Airtanker 7 (# ANG21453) of the 145th Airlift Wing, U.S. Air Force at Charlotte, North Carolina. Eight of these military Type I airtankers are available in the USA.

See also:
Fire Airbases—the airfields of the firefighters

In the early 1970s, the House of Representatives of the U.S. Congress decided to create the so-called Modular Airborne Fire Fighting System (MAFFS) to complete the existing array of state and commercial firefighting airplanes (large/heavy airtankers) for forest and surface firefighting. The eight mobile devices capable of being carried by heavy military aircraft were assigned to four units of the U.S. Air National Guard and U.S. Air Force Reserve.

The alarm system and the organizational action planning of the planes were turned over to the U.S. Forest Service (USFS) as a member of the National Interagency Fire Center (NIFC) in Boise, Idaho. There a National MAFFS Liaison Officer (MLO) takes over the coordination of action in close cooperation with the involved military agencies (units and the Pentagon).

The modular Airborne Fire Fighting System (MAFFS) consists of, in addition to the control and operating modules, a transportable tank system of five 600-gallon (ca. 270-liter) tanks with a capacity of 3000 gallons (some 11,350 liters), which was first developed and built by the Food Machinery Cor-

poration of California especially for Lockheed C-130 Hercules military transport planes. Later the devices were produced by the Aero Union Corporation, an American charter firm in Chico, California.

The complete system—compatible with the existing cargo equipment of the USAF 463L military transport—can be pushed into the hold of the airplane directly up the rear ramp of the plane from a low-loader semi-trailer without technical changes, so that through this mobility, as short a preparation time as possible can be guaranteed for the firefighting action.

After being fitted with the Modular Airborne Fire Fighting System (MAFFS), the military airplane becomes a full-fledged firefighting aircraft (large/heavy Type I airtanker).

See also:
Classification by the American Incident Command System (ICS).

As a rule, this system has been used to date exclusively in the USA or by the American forces. One exception is that Asian countries (Thailand, Indonesia), Africa (Morocco), plus Greece and Turkey, use a few of these devices. But the American MAFFS units are essentially available for use all over the world.

Most MAFFS are single-shot systems which dump their entire 3000-gallon load of extinguishing or cooling material within five seconds. They have two pipes projecting out the rear of the plane via the opened loading ramp, through which the special coolant (retardant) or water is ejected by compressed air (1200 psi/bar). A reduced dump is possible through the five-tank system.

The dropped extinguishing or cooling material forms a coated surface (fire line) with a length of about 1.6 kilometers (1 mile) and a width of about 18 meters (60 feet). The refilling of the MAFFS tanks at special fire airbases, and the release of the plane for takeoff, can generally take place in less than eight minutes. The crews of MAFFS airtankers and the ground personnel at the fire airbases are trained and retrained annually by aircraft specialists of the U.S. Forest Service (USFS).

The tanks used in the MAFFS, which have been in service for more than thirty years and are susceptible to corrosion from chemically enriched retardants, are now treated with the neutron technology of the American firm of DynCorp International LLC (DI). In cooperation with the McClellan Nuclear Radiation Center at the University of California at Davis (UCD MNRC), it is possible to use new technology to spot corroded areas in the MAFFS tanks and potential leaks. With a device resembling a medical X-ray machine, the insides of the tanks can be photographed, analyzed, and evaluated by computer.

Further Development: The Airborne Fire Fighting System (AFFS)

A new Next-Generation Roll-on/Roll-off Fire Fighting System was developed by the Aero Union Corporation—Airborne System Division (ASD). This Airborne Fire Fighting System (AFFS) replaces the Modular Airborne Fire Fighting System (MAFFS) used by the American National Guard for more than thirty years.

The new system uses modern composite materials, resulting in a reduction of overall weight, using a 4000-gallon (about 15,000 liter) tank, and is capable of reacting even more effectively to the required conditions of forest and surface firefighting through programmable drop profiles.

Thus AFFS is suitable for fighting large-scale fires from the air, up to Level 8 (the highest ICS category: 8 gallons of retardant per 100 square feet), which doubles the MAFFS achievement.

MAFFS in Action

MAFFS airtankers need an action preparation time of twelve hours for actions within the state where they are stationed (state activation) and twenty-four hours for other action within the USA (federal activation). A significant reduction of this activation time is made possible by a readiness for alarm (alert status). The alert status can be set up when the National Fire Danger Rating System (NFDRS) notes "extreme forest fire danger" and when the weather forecast gives a red-flag warning for the region.

For the alarming and forest fire action of MAFFS airtankers there are clear agreements and agreed-on procedures among the involved partners, the U.S. Air National Guard, U.S. Air Force Reserve, U.S. Forest Service (USFS), and National Interagency Fire Center (NIFC). Thus MAFFS can only be called into forest firefighting when all other suitable aircraft (airtankers) in the 50 US states are already in action or are otherwise not available for action. MAFFS are thus exclusively a supplement to state and commercial resources or a partial replacement for them.

Responsibility for action planning of MAFFS belongs primarily to the regional forestry and forest fire agencies in which a forest or surface fire is to be fought. They must be sure that regionally committed (contract) airtankers are either already in action or have to be held in readiness for a first attack. A report to that effect is sent to the National Interagency Fire Center in Boise, Idaho.

The National Incident Coordination Center (NICC) located there as a sub-department of NIFC is responsible for making sure that all nationwide contract airtankers available for forest and surface firefighting are already in action, are held as first-attack forces, or are not available for other reasons (servicing, out-time, time planning, etc.).

After that, only the Director of the U.S. Forest Service (USFS), or in his absence, his deputy (Director, Bureau of Land Management/BLM), Fire and Aviation Management (F&AM) is responsible to the National Interagency Fire Center (NIFC), is responsible for the request to the applicable military agency for the introduction of an MAFFS action.

Only the two MAFFS airtankers of the 146th Airlift Wing in Port Hueneme, California, can be put into alarm readiness (alert status) directly via the California Department of Forestry and Fire Protection (CDF).

The action of MAFFS units is linked to definite technical and organizational prerequisites. For example, the runways at the fire airbases used in the region in question must be at least 6000 feet (1.8 km) long and be paved. The U.S. Forest Service (USFS) takes over the servicing of the Lockheed C-130 on the site, and a military airbase must also be reachable within a 300-mile (ca. 480 km) radius.

Before the action, both the Lockheed C-130 and the MAFFS are checked technically and functionally. The tests are based on the time interval at which the last action of the airtanker took place. An MAFFS Liaison Officer (MLO) of the U.S. Forest Service (USFS) and an Air Force Mission Commander (AFMC) of the military cooperate closely in regulating the organizational and attack-technical action of the MAFFS airtankers.

In principle, MAFFS aircraft fly only under the direction of a lead plane, since the C-130 pilots, who are trained only in military flight-technical matters and thus are generally not familiar with the special action requirements for airtankers in fighting forest and surface fires, and thus they need tactical direction.

The year 2003 was an extreme year, not only for the firefighters in California, but also for the four MAFFS units of the U.S. Air National Guard and the U.S. Air Force. The so-called So-Cal Fires in October between Los Angeles and San Diego brought about a hitherto unknown occurrence of action for the military wildland firefighters.

In addition, the MAFFS airtankers of the 146th Airlift Wing, based at Port Hueneme, California, were called to Italy and Sardinia for the first time, where the airtankers were put to work in fighting forest fires for several weeks. The busiest fire seasons for the MAFFS units were the years 1993 and 1994, in which some 2000 flights had to be made and 25 tons of retardant dumped.

In 2006 as well, several MAFFS units saw action, mainly in forest and surface firefighting in the western states of the USA.

408
408

Firefighting Helicopters in the USA

Introduction to the Technology and Tactics of Firefighting Helicopters

Along with the airtankers (fixed-wing aircraft), the "turning-wing aircraft" (helicopters) play an important role worldwide in fighting forest and surface fires. In comparison with the numbers of airtankers, the number of fire helicopters and helitankers in use is much greater. Internationally too, most countries depend extensively on the use of helicopters in fighting forest and surface fires, sometimes in conjunction with Single Engine Air Tankers (SEAT). In addition, helicopters are used only for pure firefighting, but also as action aircraft for special teams (helitack), to transport equipment and materials to the scene of action, and as command, lead, and rescue aircraft or as observation aircraft (firewatch, infrared observation).

See also:
Information and Tips: manufacturers and contractors of firefighting aircraft (table).

Helicopters made especially for firefighting can be divided into two (or three) basic versions. In addition, in the USA they are, like the airtankers, divided into Incident Command System (ICS) Type I, II, and III categories. In non-American countries, the division generally is only that of light and large firefighting helicopters. The basic classification of firefighting helicopters includes the Fire Helicopter, the Helitanker, and the Helitack Helicopter.

Fire Helicopters (also a general term for all helicopters used in fighting forest and surface fires) are described as all helicopters that function with external load containers hung on them. These can be either the flexible and foldable Bambi buckets or rigid water containers dropped via hydraulic or radio-controlled remote control. Bambi buckets are generally filled by being dipped in bodies or basins (called pumpkins in the USA) of water, while fixed containers are generally filled from tank vehicles on the ground.

Helitankers are those helicopters that are built with rigid water containers (fixed tanks) in the underbody of the helicopter. These tanks can be filled on the ground from mobile water tenders, from hydrants, or in the air by using a loading snorkel.

Helitack Helicopters are tactically combined aircraft that give only second-tier service to the actual firefighting. They either transport a special ground crew to the scene of action, or they are available as crew helicopters to support the firefighting.

Unlike the airtankers, which are generally airplanes taken over from the military and rebuilt into firefighting planes (exceptions: SEAT, Canadair, Russian airtankers), fire helicopters and helitankers include many helicopters that were either previously in civilian use or obtained completely new. Exceptions include, among others, the older Bell helicopters and a few Type I helicopters (such as the Bell AH-1 Cobra and UH-60 Black Hawk).

In the USA, helicopters are used for forest and surface firefighting mainly by national and state forestry and forest fire agencies (including USFS, BLM, CDF) and by county and city fire departments (community and district firefighting agencies).

The same is also true of numerous other countries in Europe and elsewhere in the world, where firefighting helicopters are owned by forestry and forest fire agencies, fire departments, police organizations, rescue services, and private charter agencies.

The tactics of firefighting helicopters in forest and surface firefighting is almost identical to that of firefighting aircraft (airtankers). Unlike the much faster but less maneuverable airplanes, helicopters can be used much more precisely in terms of time and space—though of course with a much smaller load of water.

The following pages provide information on the individual functions of helicopters in fighting forest and surface fires, as well as their types and typifications.

Further information on firefighting helicopters in Europe and other countries is offered in the chapter on international firefighting aircraft.

See also:
Firefighting Airplanes/Airtankers in the USA—introduction to the technology and tactics of forest firefighting.

Most fire helicopters are equipped with flexible water containers (Bambi buckets). They are hung on the cargo hooks of the aircraft and can be filled with water in flight and dumped at the scene either mechanically or electrically. Where open bodies of water are not at hand, Bambi buckets can also be filled from water containers (like the "pumpkin" shown here).

Size Classification: Type I, II, and III Helicopters

According to the rules of the American Incident Command System (ICS), firefighting helicopters (fire helicopters, helitankers, and helitack) are divided into size groups I, II, and III. The main criterion is the amount of water, retardant or foam they can carry, plus the maximum load limit and the available passenger seats.

Accordingly, Type I helicopters (large helicopters) can generally carry more than 700 gallons (ca. 2650 liters) of liquid, and have 15 or more seats. Type II (medium) helicopters can carry between 300 and 699 gallons (ca. 1135 to 1650 liters) and have 9 to 14 seats. The lowest size class, Type III (small) helicopters, can transport 100 to 299 gallons (ca. 378 to 1135 liters) and have four to eight seats (see the table below).

See also:
Information and Tips—Aircraft for Forest and Surface Firefighting (various tables).

The greatest numbers of firefighting helicopters used worldwide are Type II and Type III models. They are used to fight forest and surface fires, usually with flexible water containers (Bambi buckets), in lesser numbers with modern fixed tanks. The number of Type I helicopters in worldwide use is considerably smaller.

Typification of Helicopters (USA) according to Incident Command System (ICS)

Type	Amount of Liquid Carried		Load Limit	Passenger Seats
	gallons	liters	lb/kg	
I	700 & more	ca. 2650 & more	5000/ca. 2268	15 & more
II	300 to 699	ca. 1135-2650	2500/ca. 1134	9 to 14
III	100 to 299	ca. 378-1135	1200/ca. 544	4 to 8

External Water Containers

The use of external water containers varies greatly in different states and in terms of realized conceptions of aerial firefighting. While Bambi buckets in different sizes and fixed tanks are used primarily on modern helicopters in the USA, Asiatic countries (Japan, China, Korea) generally use fixed tanks installed on the underside or SEMAT fixed tanks (as in Singapore and Malaysia).

Bambi buckets are preferred in Europe. Rigid external tanks are sometimes kept ready for use by helicopters in community fire departments. Leading producers of external liquid containers are, for example, SEI Industries Ltd. (Bambi Buckets®, HeliTanks™, Stilwell Flyer™) of British Columbia, Canada, and Spegel-SEMAT GmbH & Co. KG (fixed external tanks) of Augsburg, Germany. SEMAT products are used mainly in Europe, while SEI products enjoy worldwide use. Bambi buckets and fixed liquid containers are available in various sizes. The Canadian manufacturer, SEI Industries Ltd., offers Bambi buckets® in twenty sizes, from 72 to 2600 gallons (ca. 272 to 9840 liters). Some of these buckets were developed specifically for use by specific helicopters, such as the Cobra Bambi Bucket (324 gallons/125 liters) and the K-MAX Bambi Bucket (680 gallons/2590 liters).

Other products made by the SEI Industries Ltd. firm for aerial forest and surface firefighting also deserve mention here. SEI produces flexible tanks and water containers to supply water to ground and engine crews in isolated and terrain-limited hard-to-reach action areas; they can be delivered to scenes of action by helicopters as outside loads. Such external-load tanks are known as Helicopter Slung Tanks for Water Supply, Heli-Tanks®, and Helicopter Transportable Water Cells (Stilwell Flyer™), and are often known familiarly as "pumpkins." Heli-Tanks® and Stilwell Flyers™ are available with capacities from 180 gallons (681 liters) to 600 gallons (2271 liters).

The Spegel firm of Germany has developed SEMAT fixed external water containers (fixed tanks) that function according to a patented "Hubmantel" principle to release the liquid by lifting the outer wall of the container away from the base of it. The containers can be used for either quick drops (three seconds) or spray drops (15- to 30-second spray). The operation of the drop is controlled by cable or remote-control radio.

SEMAT fixed tanks are available in F and FPG versions. The F type is made for use with polymer-based liquid media and additives; the FPG also allows the use of powder and granules. SEMAT fixed tanks are available with capacities or 550 to 10,000 liters. The F 1000 is also foldable and can thus be transported in Type BK 117 and EC 135 helicopters.

SEMAT external containers have been sold in Germany (F 700, F 800, F 900, F 5000), Malaysia (FPG 900), Canada (F 700), Sweden (F 550), Switzerland (F 700, F 800, FPG 900), South Africa (F 500, F 3000), Great Britain (F 700), Slovenia (F 1800), Italy (F 800, F 900), the USA (F 5000), and Singapore (F 600, F 2000, F 3000).

The Most Important Firefighting Helicopters in the USA, with ICS Types

Model	Tactics	ICS Type	Load (gallons)
Bell 47 Soloy	Helitack	IIII	bucket 96-108
Hiller 12-D/E Soloy	Helitack	IIII	bucket 96-108
McDonald Douglas MD 500D	Helitack	III	bucket 96-108
McDonald Douglas MD 530F	Helitack	III	bucket 120-144
Bell 206 B-III JetRanger	Helitack	III	bucket 96-108
Bell 206 L-3 LongRanger III	Helitack	III	bucket 96-144
Aerospatiale AS-350 D-1 Astar	Helitack	III	bucket 108-144
Aerospatiale AS-350 B-2 Ecureuil	Helitack	III	bucket 240
Aerospatiale AS-355 F-1 Twin Star	Helitack	III	bucket 108-144
Aerospatiale SA-315B Lama	Helitack	III	bucket 180
Aerospatiale SA-316B Alouette III	Helitack	III	bucket 144
MBB BO 105 CB	Rescue	III	bucket 120
BK 117 A-4	Rescue	III	bucket 180
Bell 204B (UH-1B)	Fire/Tanker	II	bucket 240
Bell Super 204 (205)	Fire/Tanker	II	bucket 324
Bell 205 A-1 (UH-1H)	Fire/Tanker	II	bucket 324
Bell Super 205	Fire/Tanker	II	bucket 324
Bell Super-Duper 205 (Bell 212)	Fire/Tanker	II	bucket 324
Bell 212 (Bell 205, UH-1N)	Fire/Tanker	II	bucket 324
Bell 412	Fire/Tanker	II	bucket 420
Sikorsky S-58T	Fire/Tanker	II	bucket 420
Kaman H-43 Huskie	Fire/Tanker	I	bucket 324
Bell 214 B-1	Transport	I	bucket 660-800
Sikorsky UH-60 Black Hawk	Fire/Tanker	I	bucket 660
Aerospatiale AS 332L Super Puma	Fire Helicopter	I	bucket 900
Sikorsky S-61 N (Sea King)	Fire Helicopter	I	bucket 900
Boeing Vertol 107	Fire Helicopter	I	bucket 900-1000
Boeing 234 (CH-47 Chinook)	Fire Helicopter	I	bucket 3000
Sikorsky S-64 Skycrane	Helitanker	I	tank 2000
US-gallon = 3.785 liters			

Type I Helicopter 716 of the American Helicopter Transport Services, a Sikorsky S-64 Skycrane, ranks among the largest helicopters used internationally to fight forest and surface fires. It holds up to 10,000 liters of water.

An imposing view of the Skycrane in action!

The Sikorsky S-61N ranks among the large helicopters (Type I). Here we see a fire helicopter chartered by the U.S. Forest Service (USFS) (# N264F) from Pacific Aviation.

Above: Fire Helicopter 523, a Bell 212 HP (Type II) of the Kachna Aviation Corporation.

Below: Fire Helicopter 308 (# N205KS), a Bell UH-1H Iroquois (Bell 205) of the Santa Barbara County, California, Fire Department.

This Type III Bell UH-1 helicopter (# N545D) is in service with the Nevada Division of Forestry (NDF) in Carson City, Nevada.

A look into the roomy interior of a Bell 205. The firefighters' personal equipment is ready for action.

A Bell 212 fire helicopter (# N9121Z) is prepared for action. The crew hangs the water containers (Bambi buckets) under the body . . .

. . . and can pick up water from open bodies of water in flight.

N9121Z

An Aerospatiale S-315B Lama chartered by the U.S. Forest Service (USFS) at the Silver City Fire Airbase in southern New Mexico. This Type III helicopter (# N315SH) is also available for helitack action.

AEROSPATIALE
LAMA

The "Bambi Bucket" flexible water container in its transport package.

For use, it can be unfolded like an umbrella and hung on the helicopter.

A large Bambi Bucket (heavy lift bucket) has a capacity of up to 9800 liters of water.

An Aerospatiale AS-332 Super Puma of the German Border Patrol (now Federal Police) is picking up water from a mobile container.

Fire Helicopter 404, a Bell 205 (# N494DF) of the California Department of Forestry and Fire Protection (CDF, now CAL FIRE). CDF helicopters have the traditional white-red-black color layout and are thus very easy to recognize in action.

An external load container made by the German firm of Spegel-Semat. The container holds 3000 liters of water. The manufacturer's products have capacities up to 10,000 liters.

In Germany the large external-load water containers (Type II, 5500 kg) with capacities of 5000 liters can only be used by large helicopters like this Sikorsky S-65 (military CH-53-G) of the German Bundeswehr. Smaller containers (Type I, 1000 kg) with a capacity of 800 liters are suitable for helicopters of, for example, the Bell UH-1D type.

In Europe too—and worldwide—water containers are reserved for helicopter use. The picture shows a helicopter taking on water from a mobile container into an external carrier.

Twin water containers (twin heavy lift buckets, capacities up to 19,600 liters) are used in Russia. Here the crew is filling the buckets for a Mil Mi-8MTV-1 helicopter (# RA-25501) of the government forest and forest fire agency, Avialesookhrana.

A Type III Bell 206 B-III JetRanger helicopter from Riffle, Colorado, chartered by the Bureau of Land Management (BLM).

External water containers of the Balzer Fire Department near Vaduz, Liechtenstein. An Aerospatiale 5A-315B Lama helicopter (# HB-XRD) of the Rhein Helikopter AG charter firm stands ready for action.

Helitanker—The Fixed-Tank Fire Helicopter

The fire helicopters of the modern generation are called helitankers. They are equipped with rigidly built-in water or retardant tanks mounted under the cabin floor between the skids.

Fixed Tanks

They are usually filled with water but can also hold fire retardant. A foam mixture is possible depending on the technical equipment.

According to helicopter types, these tanks have capacities from 250 to 3000 gallons (ca. 950 to 11,350 liters).

While the fixed tanks of the helicopter have proved themselves in recent years in exactly directed fighting of forest and surface fires (for instance, the fighting of spot fires), not only in the USA, this system also includes a few design-related disadvantages. Problems can arise in landing the helicopter, since the free space between the tank and the ground is considerably limited, especially with uneven ground conditions. Also, the cargo hooks also located at the bottom of a helicopter are not usable with the tank in place, which is a particular disadvantage for helitack helicopters, since they are used not only for firefighting but also for transport purposes.

While fixed tanks can be removed fairly simply (it takes at most five minutes), their installation is considerably more difficult and can generally be done only with the required tool and workshop technology.

Manufacturers of Fixed Tank Systems

At this time, four makers of fixed helicopter tanks in the USA define the national and international market, while other manufacturers—which cannot be named at this point—work at a national level in various countries.

The Conair Aviation Ltd. firm of Abbotsford, Quebec, Canada, Isolair, Inc. of Troutdale, Oregon, Sheetcraft Co. of Ojai, California, and the Simplex Manufacturing Co. of Portland, Oregon, make various versions of fixed tanks for Type II and III helicopters.

Conair Aviation Ltd.

The Canadian firm of Conair has specialized superbly in making tank systems of aluminum and fiberglass components for the Bell 205A, Bell 212, Bell 206 and Eurocopter AS-350 helicopters. The tanks are fitted with two electronically operated hydraulic hatches to drop various quantities of extinguishing materials, as well as with an emergency dropping system. Loading and dropping processes can be checked on a display in the cockpit. The manufacturer's fixed tanks can optionally be fitted with an integrated foam-mixing system (exception: the Bell 205A, where the system is located in the cabin).

The quantities of materials held by the fixed tanks are, for example:

- For the Bell 205A and Bell 212, 359 gallons (1360 liters) plus 54.2 gallons (205 liters) of foam concentrate,
- For the Bell 206, 238 gallons (900 liters) plus 6.6 gallons (25 liters) of foam concentrate.

A helitanker (such as this Bell 205) differs from other firefighting helicopters by having a fixed water tank. The tank is mounted between the skids but can be removed or mounted in a few minutes if need be. Depending on the helicopter and the tank manufacturer, fixed tanks can hold up to 3000 gallons (ca. 11,350 liters) of water. Fixed tanks can be fitted with optional loading snorkels.

The California Department of Forestry and Fire Protection (CDF, now CAL FIRE) also has modern helitankers. The picture shows H-241 (# N4212Y), a Bell 205 (UH-1H) from Orange County, California.

- For the Eurocopter AS 350, 211 gallons (800 liters) plus 20 gallons (75 liters) of foam concentrate, and
- For the Eurocopter SA 315B, 238 gallons (900 liters) plus 21 gallons (80 liters) of form concentrate.

The basic version of the Conair fixed tank has a capacity of 1360 liters.

Isolair, Inc.

This firm from Oregon makes the Eliminator II fixed-tank system for the Bell 206B, Bell 206L, Bell 204, Bell 212, Bell 412, Bell 214, and Eurocopter 350 helicopters. The tanks are usually made of fiberglass. Developed from experience with the aforementioned disadvantages, Isolair has added wheels to their tanks, so that only one other person is needed to install them.

Fixed tanks made by this Oregon firm are equipped with, among other items, built-in loading pumps, two dropping hatches, foam mixers, and drop control units. The capacities of the fixed tanks are, for example:

- For the Bell 206L, 146 gallons (553 liters) plus 10 gallons (ca. 38 liters) of foam concentrate,
- For the Eurocopter 350, 198 gallons (ca. 750 liters) plus 15 gallons (ca. 57 liters) of foam concentrate,
- For the Eurocopter 350 B2, 270 gallons (102 liters) plus 15 gallons (ca. 57 liters) of foam concentrate,
- For the Bell 205, 351 gallons (ca. 1329 liters) plus 27 gallons (ca. 102 liters) of foam concentrate.

Sheetcraft Co.

The Skyhydrant fixed tanks of the California manufacturer, Sheetcraft, can be had for the Bell 205, Bell 212, Bell 206B, Bell 206L, Bell 204, and McDonnell Douglas MD 500 series. The helicopters of the Los Angeles County Fire Department (LACoFD) and the Los Angeles City Fire Department (LACFD) are equipped with Sheetcraft 205/212 tanks.

This company's tanks are made of aluminum; all tanks include a foam injection system, two drop hatches, a loading and dropping control unit, and an emergency drop apparatus. The capacities of the fixed tanks are, for example:

- For the Bell 205, Bell 212, Bell 214B, and Bell 412, 360 gallons (ca. 1363 liters), plus 12.5 gallons (ca. 43 liters) of foam concentrate,
- For the MD 500 and Bell 206B, 125 gallons (ca. 473 liters), plus 12.5 gallons (ca. 43 liters) of foam concentrate,
- For the Bell 206L, 150 gallons (ca. 568 liters), plus 12.5 gallons (ca. 43 liters) of foam concentrate.

Simplex Manufacturing Co.

This firm, active in many countries of the world (among others, Argentina, Australia, India, Japan, Korea, Malaysia, Mexico, Thailand, Turkey, South America, Europe, and the Middle East), produces Fire Attack System Tanks for Type II and III helicopters, including the following models: Bell 206L, Bell 205A1 (UH-1D), Bell 212, Bell 412 (412EP), Bell 407, Eurocopter A series (AS 332, AS 350, AS 355, AS 365), BK-117 Kawasaki). In addition, fixed tanks are also available for Type I helicopters: Sikorsky S-64 (CH-54), Eurocopter AS 332 Super-Puma, Agusta A 119, and Kamov KA-32.

One of the largest helitankers is the Sikorsky S-64 Skycrane. H-72 (# N217AC) is a Type I helicopter of the Erickson Air-Crane firm, equipped with a 2000-gallon (ca. 7570-liter) tank, a loading snorkel, and a water cannon. Erickson Air-Crane is one of the most important international charter firms for heavy load-carrying and forest firefighting helicopters.

The tanks built by this firm are made of fiberglass and can carry either water, fire retardant, or a water/form mixture. Special foam-concentrate tanks are built into the tanks. The pilots have control devices for loading and controlled dropping (loading quantities, foam mixing, drop control, action documentation). All systems are operated electrically or hydraulically. The capacities of the fixed tanks are, for example:

- For the Bell 206L (L1, L3), 155 gallons (ca. 587 liters), plus 14 gallons (ca. 53 liters) of foam concentrate,
- For the Bell 205 and 212, 360 gallons (ca. 1363 liters), plus 25 gallons (ca. 95 liters) of foam concentrate,
- For the Eurocopter AS 350B (C, D), 155 gallons (ca. 587 liters), plus 20 gallons (ca. 76 liters) of foam concentrate,
- For the Eurocopter BK-117, 211 gallons (ca. 800 liters), plus 14 to 20 gallons (ca. 53 to 76 liters) of foam concentrate.

Badger Fire Suppression Tanker System

The newest developments allow the filling of fixed tanks—in addition to filling on the ground through the use of water tenders and hydrants—via snorkel loading. Appropriate equipment is produced by, for example, the Tower Aerospace firm of Calgary, Alberta, Canada. The T-5000 Otter Linear Delivery System, for example, an electronically and computer-controlled fixed tank with a loading snorkel, can hold, depending on the helicopter model, up to 1000 gallons (ca. 3785 liters) of liquid. A foam mixing system (3% mix) is integrated in the tank. The complete loading process in flight via the loading snorkel takes between 15 and 45 seconds (300 gallons in 15 seconds, 500 gallons in 22 seconds, 100 gallons in 30 to 40 seconds).

The system can be operated by a minimum of one person, the pilot.

CDF Helitanker 303, a Bell B-212 (# N500EH), from San Diego County, California.

Bell B-205A-1 (UH-1H, # N503SH), of the American Silver State Helicopters charter firm of North Las Vegas, Nevada, is making a drop from a fixed tank. The copter is fitted with a Simplex Fire Attack tank holding 369 gallons (ca. 1397 liters), and 30 gallons (ca. 113 liters) of foam concentrate.

Sikorsky S-64 Skycrane H-794 (# N44094), of Heavy Lift Helicopters, in flight with its loading snorkel down.

Another Helitanker from Erickson Air-Crane, H-746 (# N223AC), is taking on water from a pond.

CDF Helitanker, Bell B-212 (# N509EH), equipped with fixed tanks and loading snorkel. The pictures show in detail the construction, attachments, connections, and transport rollers of such tanks for the undersides of helicopters.

AIR OPERATIONS
BARTON HELIPORT

Helitanker 11, Bell 412, of the Los Angeles County Fire Department (LACoFD, # N412LA) is equipped with a Skyhydrant 205/212 tank (360 gallons, ca. 1363 liters) made by the Sheetcraft firm of California. This helitanker is also equipped with a cable winch and used for rescue and retrieval work.

Helitanker 12 (# N12OLA), another Bell B-412 of the Los Angeles County Fire Department (LACoFD).

Helitanker 16 (# N160LA), a Bell B-205A-1 of the Los Angeles County Fire Department (LACoFD).

Helitack—
The Helicopter of the Rappelling Crews

Helitack helicopters have a double function, and not just in the USA. They are used both to transport members of the so-called helitack or rappelling crews (*), and needed equipment and materials (personal equipment, tools, provisions) to the scene, as well as to fight fires.

Helitack helicopters are firmly assigned to the helitack crews established in the readiness areas. They are usually Type II (medium) or Type III (small) helicopters, which are normally equipped with flexible external water containers (Bambi buckets). Non-American countries (such as Russia) sometimes use large (Type I) helicopters for numerically larger crews of firefighters.

Helitack crews are specially trained units (of 10 to 15 firefighters) who are brought to scenes of action in helicopters. They are not to be confused with the special units of smokejumpers. Helitack crews, though, like smokejumpers, are classed as initial action crews where action from other ground crews is not possible in the required time (for example, because of too-long approach times or inaccessible terrain).

Special crews of this kind are used mainly to fight fires in the USA, Canada, Russia, and some Asian countries (such as Japan).

A special type of helitack crew has been used by the California Department of Forestry and Fire Protection (CDF) for some time, and is known within the realm of aerial firefighting as "Short Haul Rescue." The crews are specially trained rescue crews who are generally used to rescue people from areas endangered by forest fires (for example, buildings, roofs, landscapes). All helitack crews of the CDF (180 members in all) have also been trained as "Short Haul" rescuers. Special units of this type are also found in Asian countries.

At the scene of action, the crew members are set on the ground directly by the helicopter or lowered to the ground by rope. There the helitack crew begins the firefightng measures as a ground crew, and they can later be supported by the helitack copter with water dropped from the air (Bambi buckets).

But the helitack copter can also be used to bring additional crew members or additional equipment and materials. For tactical reasons, this helicopter is not intended to used as a primary fire helicopter.

As for the CS classification and the technical equipment, the same standards apply to them as to Type II and III fire helicopters (and in exceptional cases, Type I copters).

Yet helitack copters are not usually fitted with fixed tanks. For one thing, the cargo hooks attached to the underside need to be usable at all times, and for another, landing problems due to bottom-mounted tanks should be avoided.

(*Rappelling refers to a descent and landing by roping technique from a flying helicopter using a rope and personal roping equipment.)

The Kern County Helitack crew (Kern County Fire Department) uses H-408, a properly equipped Bell 205A-1 (Bell UH-1H, # N408KC). The crew, under the command of Helitack Captain Kevin V. Loomis (front, third from left), is stationed at Keene, California.

MINUTEMAN
N53MA

Helitack in action! Helitack crews and all their equipment are unloaded from the helicopter at the scene of action or—if they cannot land—lowered to the ground by rope. The helicopter (here a Bell 206B-3, # N53MA, Minuteman Aviation), then can either bring additional personnel or materials and equipment, or take on firefighting tasks using its Bambi buckets.

The Russian national forest and forest fire agency, Avialesookhrana, also uses helitack crews in forest firefighting. The pictures show a crew being lowered from a Mil Mi-8MTV-1 helicopter (# RA-25501).

Members of a helitack crew of the Bureau of Land Management (BLM) are roping down from a Bell 205A01 (# N214R).

Helitack in action!

The Helitack Crew 555 of the Bureau of Land Management (BLM) is stationed east of Bakersfield before Kern Canyon. After fighting a fire, they can relax under the California sun.

During his many visits to wildland stations in the USA, Author Wolfgang Jendsch (4th from left) had several opportunities to fly with Helitack Crew 408.

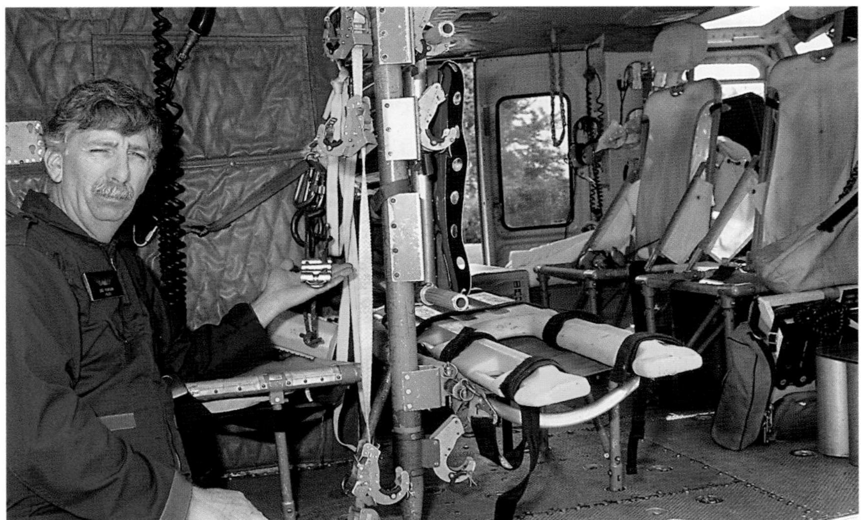

The crew of Helicopter 11 of the Los Angeles County Fire Department (LACoFD) with Senior Pilot V. Lee Benson (2nd from left). The LACoFD helicopters are intended not only for fighting fires, but also, because of their special equipment (such as rescue and recovery equipment, ropes, and cranes), also for helitack and rescue action.

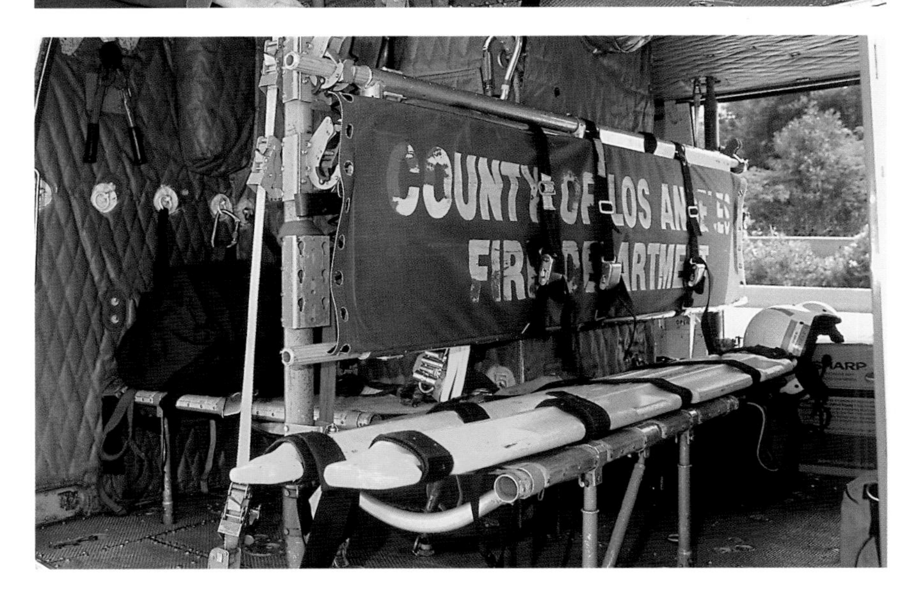

Helitack 555, a Bell 212 used by the Bureau of Land Management 9BLM) (# N212KA), fighting a fire after landing the crew.

Helitack 408 is in action with its crew over Lake Isabella, not far from the Sequoia National Forest in California.

Among other things, helitack crews are equipped with backpacks with personal items, drinking-water bottles, fire shelters (rescue tents), and tools (including McLeods, Pulaskis, and chain saws). Ropes and roping materials are also carried on the helicopter.

MEADOWS FIELD

Helitack Helicopter 530—Bell 206L-3 LongRanger III (# N383SH), from Aspen Helicopters, Inc., is chartered by the U.S. Forest Service for forest firefighting in Los Padres National Forest.

Helitack Helicopter 552, an Aerospatiale AS-350 B2 Ecureuil (# N505WW), of Rogers Helicopters, Inc., is used by the National Park Service at Kings Canyon.

A Bell 407 helitack helicopter.

A Bell 206 B-3 JetRanger helitack helicopter (# N296N) is seen at a helispot not far from the Williams Incident (September 1997) in Yuba City, California.

Helitack Helicopter 52, a Bell 212 (UH-1N, # N73HJ) from Sequoia National Forest (Peppermint Base), chartered by the U.S. Forest Service (USFS).

Special Airplanes for Fighting Forest Fires in the USA

Command Planes

Firefighting must be led and directed—this is not only the practice in the USA, but in almost all the countries of the world. And what applies to leading the action of ground crews is not essentially different in fighting forest and surface fires from the air.

The American Incident Command System (ICS, Operations Section—Air Operations Branch) provides for an Air Commander or Air Tactical Supervisor for action that requires air support. He coordinates the air attack (airtankers, helicopters, jump planes, transport planes) that operates over the scene of the action in close cooperation with the ground commander or via the liaison person with the General Incident Commander.

During an action, and according to tactical and/or informal requirements, the air commander can circle the scene for hours with his command plane, so as to make the required tactical decisions for the air attack operations.

The air commander (or air attack supervisor) has an air command plane or command plane. This is usually a roomy light or medium-sized passenger and/or transport plane that is supplied with the necessary equipment for directing the action (such as radio, GPS, GIS, EDV, and work-space).

Among others, the following types of airplanes are usually used as command planes:

- Aero Commander 500B
- Beechcraft V35B Bonanza
- Cessna 206 Super Skylane
- Cessna 337 Skymaster
- Turbo Commander 690

Tactical meeting: Airtanker 26 along with a command plane of the U.S. Forest Service.

Helicopters are rarely used as command planes, or only in limited instances. What with their speed, their flying characteristics, and their space, they are usually unsuited or only slightly suited for this task. During the action flight, the air commander has not only the pilots but also one or more assistants for observation, radio communication (dispatchers), or documentation. If need be, other persons can accompany the flight—for example, additional leadership personnel of the forestry and forest fire agencies, specialists or representatives of fire departments.

The Kern County Fire Department (KCFD) has an Aero 500B Commander (AA 490, # N911KC) as its air attack plane/command plane. The plane is used as a flying air commander in forest and surface fires.

Command planes are usually stationed at the fire airbases but can also be kept at regular airports. Often the locality of the regular workplace of the responsible air commander makes the difference. These command planes may be owned by either the forestry and forest fire agencies or the larger fire departments.

The tactical concept of air attack operations is dominant in the USA. In Europe and other countries the function of the air commander is relatively unknown in this form. Aircraft used in fighting forest and surface fires are generally directed there from the ground.

An example of this is the "Officer Aero" (flight-leading officer) of the French firefighters. Years ago, after negative experiences with leading the action of ground and air units under a joint command, the French introduced the position of "Officer Aero" so as to improve the qualification and safety of the air attack by separating the two functions, and at the same time unburdening the chief action commander.

The Officer Aero fulfills a ground-bound function on the principle that the action leadership is directly divided. In larger actions, the Officer Aero usually has an aircraft (more often than not a helicopter) at his disposal for information and coordination flights over the area.

A Beechcraft King Air 100 command plane.

Below: The graphics show the work and functioning areas for aircraft in fighting forest fires.

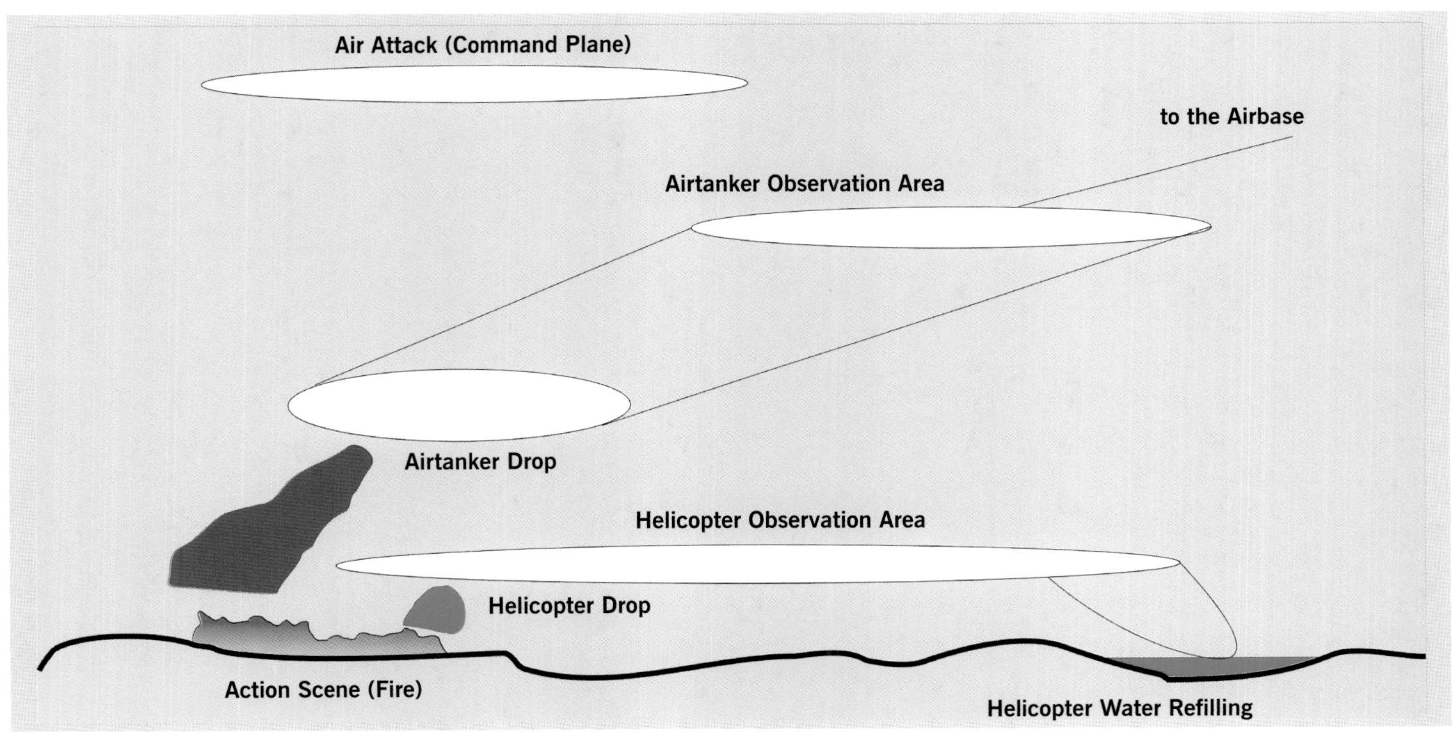

Command Plane, Air Attack 01, # N1176Z, an Aero 500B Commander of the U.S. Forest Service (USFS). The plane is in service at the Angeles National Forest.

A Bell 206 B-3 JetRanger helitack helicopter (# N296N) is seen at a helispot not far from the Williams Incident (September 1997) in Yuba City, California.

Helitack Helicopter 52, a Bell 212 (UH-1N, # N73HJ) from Sequoia National Forest (Peppermint Base), chartered by the U.S. Forest Service (USFS).

Command Plane, Air Attack 15, # N24GT, an Aero 500B Commander of the U.S. Forest Service (USFS), stationed at the Fresno Air Attack Base.

Command Plane, Lead Bravo 5, # N7059H, Dynamic Avlease Inc., a Beech 65-A90-1 King Air, used by the U.S. Forest Service (USFS).

Lead Planes

See also:
Graphics, page 187—work and functioning areas.

Lead planes could really be described as the "follow me" vehicles of the air. As on the ground, in the area of an airport, these vehicles guide incoming aircraft to their assigned places, the function of the lead plane in aerial firefighting in the USA can be seen in a similar light.

These usually fast and nimble propeller planes meet the firefighting planes (airtankers) at the fire airbase and conduct them from there to the action area. Once arriving at the action area, the planes are assigned either to a holding orbit at first, or directly to a drop orbit for retardant (see also the "Air Attack Operations" graphics in the "Command Planes" chapter).

After the drop, the airtankers may be led back to the fire airbase by the lead plane.

One of the most modern lead planes used by the California Department of Forestry and Fire Protection (CDF, now CAL FIRE) is this Rockwell OV-10A Bronco (Lead 430 Fresno), # N407DF. The CDF has rebuilt 13 of these ex-military planes and put them into service for leading tasks in the air.

Airtankers are used to fight forest and surface fires in the entire area of the ten western states of the United States, and in some cases throughout the USA. Thus, as a rule, the crews of the planes are not familiar with the localities. Even air maps on board and the Global Positioning System (GPS) do not make the location of action scenes and definite areas of fire zones much easier in the incredible vastness of the often-unbroken landscapes and wildlands.

Being led directly to the action and drop site by a lead plane has proved to be tactically the fastest and surest method in aerial firefighting.

The use of lead planes for firefighting planes going into action is limited to the USA. Comparable conceptual and tactical systems from other countries are not known.

In the USA, only a few types of planes are available for these tasks. They are, among others:

- Beechcraft V35B Bonanza
- Beechcraft 55 (58, 58P) Baron
- Cessna 205 (206, 206A)
- Cessna 337 Skymaster
- Rockwell OV-10 Bronco

With thirteen Rockwell OV-10 Broncos, former U.S. Navy and U.S. Marine Corps planes, the California Department of Forestry and Fire Protection has the largest and most modern fleet of lead planes for use in fighting forest and surface fires.

A look into the narrow cockpit of the OV-10A Bronco, with the pilot's seat in front and a workplace behind it for a navigator or air commander.

Linda Scott of the Fresno Airtanker Base (CDF/USFS), one of the few women lead-plane pilots in the American wildfire community.

A lead plane of the U.S. Forest Service (USFS), a Beech 58P Baron (# N165Z).

An identical lead plane of this type stands at the airport of the National Interagency Fire Center (NIFC) in Boise, Idaho. The U.S. Forest Service (USFS) has numerous planes of this type in use.

A Beechcraft 58P Baron lead plane of the U.S. Forest Service (USFS) in action.

Lead-plane pilot and "fire-pirate" Michael (Mike) Lynn (right) from Lancaster, California.

A Beech Super King Air 200 (# N200EJ) lead plane.

An often-used type of plane becomes history: These Beech 58P Baron lead planes of the U.S. Forest Service (USFS) are about to be mustered out.

Transport Aircraft for Smokejumpers (Jump Planes)

Smokejumpers, who are sometimes described in Europe as "death-defying fire-jumpers," are specially trained forest firefighters, usually with years of experience. They go into action via parachute jumps from an airplane, and usually land very close to the heart of the fire.

Their evaluation as "death-defying fire jumpers" is basically false, for smokejumpers are regarded as initial-action forces in places where the beginning of firefighting is not possible because of problematic terrain conditions and/or the inability of ground crews to reach the area within a reasonable time. Their action is technically and tactically most highly prepared, calculated, and basically safe. They are not death-defying, nor do they jump into a fire—for smokejumpers usually land in a safe surrounding area of a forest or surface fire and sometimes must walk long distances to reach the actual fire.

After they begin firefighting activity on the ground, smokejumpers generally stay in the area only 48 hours at most, until ground forces (such as hotshot crews and engine crews) have arrived and taken over the firefighting work.

Smokejumper crews are composed of groups of 15 to 20 men and women. They are either professionals or seasonal volunteers in the service of the large forestry and forest fire agencies (USFS, BLM, interagency units).

Among the best-known team and transport airplanes (jump planes) used by American smokejumpers is the Shorts SD 330-200 (C-23A Sherpa in the military). The picture shows the plane of the BLM smokejumpers from Boise, Idaho (# N188LM). The jump plane is laid out for ten jumpers, other personnel (spotters), and their complete equipment.

In the USA there are six smokejumper bases with some 440 persons. Of them, the Bureau of Land Management (BLM) has two units with some 130 members and four jump planes (three DHC-6 Twin Otter BLM and Leading Edge, # N49SJ, N97AR, and N252SA, and one Fairchild Dornier 28, Bighorn Airways, # N266MC).

The other four units are managed by the U.S. Forest Service (USFS). They use Douglas DC-3, DHC-6 Twin Otter, C-23A Sherpa, C-212 Casa and Fairchild Dornier 228 jump planes:

- Smokejumper Unit Great Basin (BLM) in Boise, Idaho (NIFC, 64 members, 3 jump planes),
- Smokejumper Unit Alaska (BLM) in Wainwright, Alaska (68 members, five jump planes),
- Smokejumper Unit Region 1 (USFS), in Missoula, Montana (70 members, 2 jump planes), in Grangeville, Idaho (29 members, one jump plane) and West Yellowstone, Idaho (20 members, one jump plane),
- Smokejumper Unit Region 4 (USFS) in McCall, Idaho (70 members, two jump planes),
- Smokejumper Unit Region 5 (USFS) in Redding, California (40 members, two jump planes),
- Smokejumper Unit Region 6 (USFS) North Cascade (24 members, one jump plane), and in Redmond, Oregon (35 members, one jump plane).

Information and Tips—Table 2, aircraft for forest firefighting—technical data.

During the annual fire season, smokejumpers can be stationed temporarily at other fire airbases (for example, Silver City Fire airbase or Cedar City Fire Airbase, see text). Every smokejumper unit has at least one so-called jump plane as a team and transport aircraft. Depending on the type, such a plane can carry from six to twenty men with all their equipment (up to 175 pounds, ca. 80 kg).

Also among the crew of a jump plane are the pilot and one or two spotters, who are responsible for the safety and coordination of the jump.

Much greater than those of the USA are the technical and personal resources in the Russian states. There some 4500 smokejumpers and numerous Antonov AN-2 jump planes are on hand to fight forest and surface fires.

АВИАЛЕСООХРАНА
RA-33370

Smokejumpers jump from an Antonov AN-2 (# RA-35171) of the Russian national forestry and forest fire agency, Avialesookhrana. During the forest fire season the agency has some 4500 smokejumpers ready for service.

Twin Otter (DHC-6, # C-GKBC) at the airport of the National Interagency Fire Center (NIFC) in Boise, Idaho.

Below: Shorts C-23A Sherpa (# N179Z) of the U.S. Forest Service (USFS), stationed at the Redding Airtanker Base.

A USFS smokejumper after the jump. Among his personal equipment are, in particular, his protective clothing (jump suit), helmet (hard hat), specially designed and secured jump boots—and naturally the parachute.

An American smokejumper of the U.S. Forest Service (USFS) from Missoula, Montana. An interesting detail: USFS smokejumpers usually jump with round FS-14 parachutes, while BLM smokejumpers prefer rectangular Ram Air Square parachutes. The reason for this is the different terrain: The U.S. Forest Service works mainly in forests, while the Bureau of Land Management (BLM) also deals with flat wildlands and agricultural lands (meadows, grasslands). Strong winds and gusts are met there, and in such cases the Ram Air Square parachutes have proved themselves especially well.

Boise smokejumpers before going into action, cooperatively checking each other's equipment goes without saying here and can save lives.

The interior of a USFS jump plane; at right is the bench for the ten to twelve smokejumpers, at left sits the spotter, behind him is space for tools and equipment. The pilot's seat can be seen at the front.

A female smokejumper ready for action, with jump suit, hard hat, parachute, and a small backpack of personal equipment.

At the Redding Fire Airbase in California a Cessna 208 Grand Caravan jump plane (# N208N) is stationed. This is the 12.67-meter lengthened version of the Cessna 208 Caravan (11.46 meters), which has room for six smokejumpers including all their equipment.

De Havilland Canada DHC-6 Twin Otter (Grand Canyon Airlines, # N171GC), used as a jump plane by the U.S. Forest Service (USFS), seen here at the Cedar City Fire Airbase in Utah. Lower right: The smokejumper crew of the Cedar City Airtanker Base line up before their plane.

N208N

NORTHWEST IOWA
TECHNICAL COLLEGE

A DHC-6-300 Twin Otter jump plane (De Havilland-Canada, # N302EH) of Era Aviation, designed for eight to ten smokejumpers with their equipment.

Right page: A look into the cockpit of the DHC-6 Twin Otter.

Other necessary equipment and materials are also carried in the jump plane and dropped on the scene by parachute.

FLAPS
FRICTION
FRICTION
THROTTLE
PROP
FUEL
OFF
WINDSHIELD
PROP AUTOFEATHER
N171GC
CAUTION

An especially old but "dearly loved treasure" of the smokejumpers is this Douglas DC-3 Turbine (# N115Z) of the U.S. Forest Service (USFS), dating from 1944. This roomy jump plane can carry twelve to twenty smokejumpers and their equipment. The picture shows the plane, later fitted with turboprop engines, that won fame in Germany as the "Raisin Bomber" after World War II, before a mission at the Silver City Airtanker Base in New Mexico.

Below: The Boeing KV-107 Vertol (Kawasaki, # N185CH) of Columbia Helicopters Inc. of Portland, Oregon, is used as both a fire helicopter and a jump plane.

The interior of the Douglas DC-3, with room for personnel and material.

International Firefighting Planes: Europe and the World

International Forest and Surface Firefighting—Technology, Tactics, and Conceptions

See also:
Firefighting planes/ Airtankers in the USA—The Single Engine Air Tanker (SEAT).

See also:
Water Bombers (Scoopers—Amphibian Airplanes).

The technical and tactical exposition and most of the photographic material in this book on the fighting of forest and surface fires in the world are based—as mentioned many times already—primarily on the practical concepts of aerial firefighting in the United States of America (USA). Except for a few special concepts (such as in the Russian states, some southern and eastern European countries or in Asia, where they have had their own concepts from the start—though they are sometimes very similar to American aerial firefighting concepts), the practical experience and time-tested technology and tactics of the USA are used almost worldwide. For example, in non-American countries there large airtankers are often used rather than smaller planes (such as the agricultural tanker or Single Engine Air Tanker/SEAT). Further, the tactical cooperative involvement of airtankers and fire helicopters is practiced in almost all the world.

Coming not from the United States, but from nearby Canada, a different firefighting concept has found enormously widespread use, especially in Europe, but very probably worldwide as well: the concept of the Canadair amphibian airtanker made by the Bombardier firm of Canada.

In most cases, US concepts were and are certainly not adopted "one-on-one" in other countries, but rather are adjusted to suit the country's own internal and regional conditions and needs. So it is not by chance that American wildland fire managers are particularly in demand at international specialist conferences and training symposia, and that in long-term large catastrophes (such as forest fires in Australia) American wildland firefighters and air attacks are called on for support, or American firefighting aircraft and/or equivalent technical equipment find ever-growing markets in Europe and other countries. One of the best examples of this is the Texas Air Tractor firm, which sells its Single Engine Air Tanker (SEAT) successfully in Europe and other non-American countries.

Land-Based Concepts and Techniques of Forest and Surface Firefighting

At this point it is, of course, not possible to assemble and publish all the national, regional or organizational concepts used in the world for forest and surface firefighting from the air. In the end, this is also not the purpose of this book. But if several such concepts can be examined and presented in brief, then they can serve as examples for many other similar concepts that are used in various lands and landscapes.

If one looks at basic concepts of aerial firefighting, then one sees four inherently different systems that are used internationally to fight forest and surface fires, or systems that vary in certain details. The following division into main points does not judge the firefighting systems qualitatively, but simply strives to make the different concepts of cooperation between ground and air units clear.

Previous double page: In France, Dash DH8-Q400 (Bombardier) firefighting planes of the Securite Civile come into action. Plane 73 (Milan 73, # F-ZBMC) is stationed at the fire airbase in Marignane near Marseilles.

This Eurocopter AS-350B-1 Ecureuil is in action in a forest fire in the Canadian province of Ontario. At this time, nine Canadair CL-415 and five DHC-6 (de Havilland) Twin Otter planes are in service in Ontario, plus large numbers of fire helicopters and helitankers.

In addition to the airtankers include fire helicopters to the French concept of fighting forest fires from the air. Pictured is a helitanker of the type Aerospatiale AS-350B-1 Ecureuil (Morane 1) at the Gare d'Aix TGV. Forest fire on July 11th 2004.

Along with firefighting planes, fire helicopters are part of the French concept of fighting forest fires from the air. Here an Aerospatiale AS-350B-1 Eucreuil (Morane 1) helitanker of Gare TGV of Aix fights a forest fire on July 11, 2004.

• The American system of aerial firefighting, which is undeniably one of the most successful in the world to date, is presently undergoing a serious change. It is designed according to the Incident Command System (ICS), which foresees close tactical and technical cooperation of high-performance ground units and a combination of air forces including airtankers, fire helicopters, special air units (helitack, smokejumpers), and command units (air command, lead planes, infrared documentation).

• The Russian system, very similar to the American in parts through its history, but proceeding from an extensively independent parallel development. Here too, fire airplanes of different sizes, fire helicopters, helitack, and smokejumper units all see service. The emphasis on effective and high-performance ground units is less pronounced, as is an all-encompassing cooperative command structure.

• The southern European system of aerial forest and surface firefighting used by most of the Mediterranean countries is based on a widespread use of amphibian firefighting planes (Canadair airtankers) in cooperation with fire helicopters, helitankers). Helitank, and smokejumping in particular, are used only in individual cases or not at all. In many cases, an equal performance potential of air and ground forces does not exist.

In Eastern Europe in particular, police helicopters (like this Bell 412 HP of the Czech Policie) and the military are used to oversee the forests and fight forest and surface fires.

Croatia: A Canadair CL-415 amphibian firefighting planes drops water on a fire.

The Russian system of aerial support in fighting forest and surface fires—similarly to that of the USA—is based on a tactical combination of large airtankers, helitankers, helitack copters, and smokejumper units. This picture shows an Antonov An-26 (# RA-26002) firefighting plane at the Russian national forestry and forest fire agency, Avialesookhrana.

Along with the big amphibian airplanes, the southern European lands use smaller Single Engine Air Tankers (SEAT). This Air Tractor AT-802 Fire Boss can also be used with pontoons.

The charter firm of Era Aviation uses, among others, this Bell 412 (# N169EH) at the Centro Forestale of the Sierra de Cazorla national park in Andalusia, Spain.

In most countries around the world, aerial fighting of forest and surface fires is based on a concept involving forestry and firefighting agencies, fire departments, charter companies that specialize in firefighting, and the military and police. Organizations' own and chartered fire helicopters and helitankers are used. In many countries, aerial firefighting—along with aerial rescue service, mountain and water rescue or searching—is simply a part of the fire department's tasks.

But there are also countries that seem to get by without a special overall concept of aerial forest and surface firefighting. "Seem," perhaps, because there are naturally forest and surface fires to be fought in these countries, but as a rule they are generally fought and extinguished only from the ground without any (significant) support from aircraft. Generally, large-volume tank vehicles are used in great numbers, and if water from the air proves to be needed in the end, police, military or fire department rescue organizations provide support with their helicopters.

Germany

Germany is very typical of this concept. Here there are neither firefighting airplanes nor specialized firefighting units for fighting forest and surface fires, which, in fact, do not occur all too often.

Existing regional and/or statewide concepts generally concern firefighting from the ground. In some cases, helicopters of the Federal Police (formerly the border patrol), State Police or the Bundeswehr, and in a few cases the rescue helicopters of the help organizations (such as the Allgemeiner Deutscher Automobilclub/ADAC or the Deutsche Rettungsflugwacht/DRF) are called on for air support. A few private charter organizations also make helicopters and/or airplanes (SEAT) usually used in agriculture available—but in many cases, sometimes insuperable bureaucratic problems must be included in the federal or existing firefighting concept (for example, in catastrophe protection or in the framework of state firefighting laws).

Attempts to develop concepts for modifying airplanes and/or helicopters for firefighting (conversion) have had little success to date. The equipping of a military Transall C-160 with water tanks around 1968-70 was a single undertaking, and the appropriate sets of equipment are either not in use now or are long since non-existent.

Considerations of private economic undertakings to equip or re-equip large military transport planes or develop new ones (such as the Airbus A 400 M as successor to the Transall C-160) remained, at least to date, merely concepts on paper, and consideration of a large service undertaking to convert Bell military helicopters to helitankers remain at present only a "specialized theory" or are still in the planning stages.

A Sikorsky CH-53 G medium transport helicopter of the German Army.

The German Border Patrol (now Bundespolizei) and three of their SA-330J Puma copters helped to fight forest fires in Portugal in July 2003. They flew with Bambi buckets of up to 2000 liters of water. The picture shows one of the crews preparing for action. The Bundespolizei (Federal Police) is also available in Germany for air support in forest and surface fires when needed.

At this time there are only a few aircraft that can be used by German firefighters to fight forest and surface fires. In the East German states, there are still a few agricultural airplanes left from DDR days and owned by private contractors—such as the PZL M-18 Dromader. From former DDR supplies there also remains the Czech Zlin 37, likewise an agricultural plane, and a few Russian-made helicopters.

The Bundespolizei and Bundeswehr (armed forces) presently use, among others, their AS 332 Super Puma (Eurocopter), CH-53 (Sikorsky S-65), and Bell UH-1D helicopters for firefighting. They are powerful enough to be able to carry external water containers (as a rule, the 900-Liter "Smokey III" and 5000-liter "Smokey I") that are reserved for use by the firefighters.

Lighter helicopters, such as those of the state police and rescue organizations, can operate only with Bambi buckets.

In addition, during the annual forest fire season, honorary firefighters, volunteer sport flyers, are ready to use their private planes to help patrol fire-threatened forests in some German states (such as Bavaria, Niedersachsen, Rheinland-Pfalz, and Baden-Wuerttemberg) in cooperation with the local

The German state police organizations also use their helicopters for firefighting. In the picture is a PZL W-3 Sokol (# D-HSNA) of the Saxon Police helicopter squadron. It is equipped with a Bambi bucket.

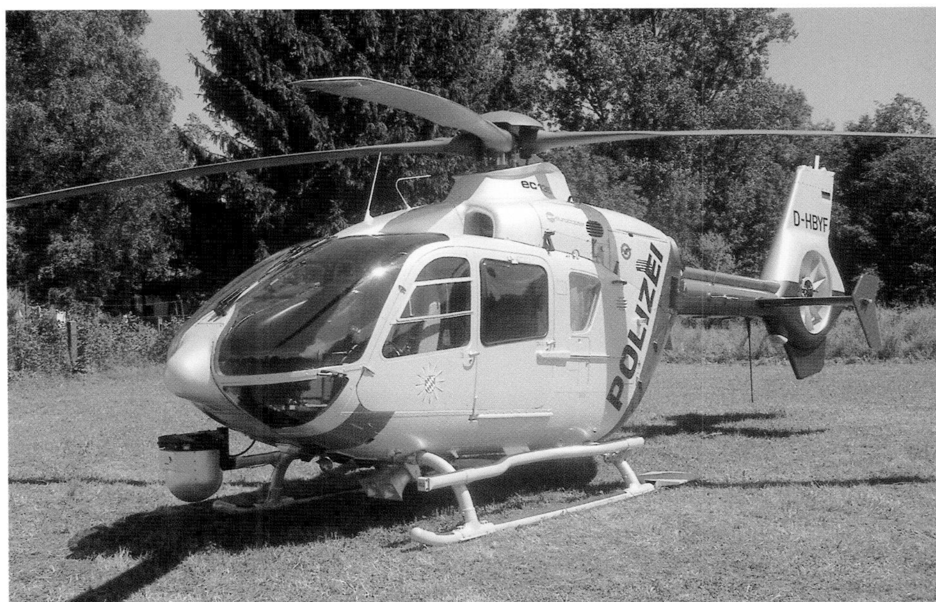

The Eurocopter EC-135P-1 (# D-HBYF) of the Bavarian Police helicopter squadron ranks among the most modern copters of the German police. An interesting detail: The copter is fitted with a camera (at the front of the right skid) that can transmit its pictures directly to an action headquarters—a valuable support tool, especially in large forest and surface fires.

forestry officials and fire departments—for example the Luftrettungsstaffel Bayern e.V. (German Aerial Observation Service) and the Feuerwehrflugdienst of the state fire agency of Niedersachsen. They use, among others, various small planes like the Cessna 172, Cessna 182 or Zlin 143, plus some helicopters.

The rescue helicopters of the Allgemeiner Deutscher Automobil-Club (ADAC) are also suitable for firefighting. This picture shows the rescue helicopter "Christoph 72" (# D-HBKK) stationed in Murnau, Bavaria.

Intensive Transport Rescue Helicopters (ITH) of the Deutsche Rettungsflugwacht (DRF, Bjorn Steiger Rescue Society) include this Eurocopter-Kawasaki BK-117 B-2 (# D-HUUU), still in its old colors, stationed in Freiburg.

A flight over the Black Forest to Freiburg in Breisgau.

ACHTUNG
NUR VON
VORNE
HERANTRETEN
K-MAX
D-HFZA
ZAGEL
HELI AIR

Among the German charter companies that have helicopters ready for firefighting is the Heli Air Zagel (now HELOG Lufttransport KG). One of their most modern copters—along with the Kamov Ka-32, Aerospatiale AS-332 Super Puma, Mil Mi-26, Aerospatiale AS-350 B3 and Aerospatiale SA-315B Lama—is the Kaman K-1200 K-Max (# D-HFZA).

The Bundeswehr's Bell UH-1D is found very often fighting forest fires in Germany. The copters—seen here with one of the crews—are reserved as SAR (Search and Rescue) craft and partially involved in rescue service as well.

D-HOXP
BERLINER SPEZIAL

The Berliner Spezialflug GmbH—BSF Hubschrauber-Technik-GmbH (BSF, no longer in existence) also made their Russian Mil Mi-8T transport and freight helicopters available on request for firefighting. This copter (# D-HOXP, once part of Interflug, DDR) was stationed at the Berlin-Schoenefeld Airport.

Lower left: A Bundeswehr Transall C-160 was used in the late sixties and seventies to introduce the large firefighting airplane in Germany. The efforts were not followed up.

The Feuerwehrflugdienst was or is organized in some German states, mainly by private aviators. Along with members of the forest service, observation flights were made over fire-threatened regions of Germany, so as to help spot forest fires as soon as possible. The picture shows a Cessna/Reims 172 (# D-EBAG) used in Niedersachsen (Lueneburg District) during a program put on by the state fire protection organization of Niedersachsen at a firefighting fair.

This Aerospatiale AS-332 Super Puma of the Border Patrol (now Bundespolizei) is on its way to a fire (seen in the background).

Austria

Since 1978, all the Austrian states have set up special support points of the fire department (air service) for forest firefighting (for example, Flugdienst Steiermark, with 66 members, Flugdienst Niederoesterreich, etc.). Equipment and materials for aerial firefighting are kept there; volunteer firemen (including flight helpers, observers, action leaders, and aerial rescuers) are available.

To fight forest and surface fires from the air, surface airplanes and helicopters, including those of the Austrian Army (Pilatus PC-6 B2H2 Turbo Porter/SEAT, Agusta-Bell AB-212, Bell UH-1D, Agusta-Bell B-204, Eurocopter AS-350 B/B1 Ecureuil, Sikorsky SA-70A Black Hawk, McDonnell Douglas MD 500E), of the Austrian Ministry of the Interior (including Eurocopter AS-350 B/B1 Ecureuil, Eurocopter AS-355 F2 Ecureuil, Eurocopter AS-355 N Ecureuil), and in special cases those of air rescue services (such as ÖAMTC Eurocopter EC-135) are used.

Switzerland

There is no federal system of air-supported forest and surface firefighting in Switzerland. Nor are there any special aircraft available for fighting forest fires. If necessary, the fire departments call particularly on the various charter companies (such as Helog, Air Zermatt, Air Grischa), and especially on the helicopters of the mountain rescue organizations (including the Schweizerische Rettungsflugwacht, Rega, Air Zermatt).

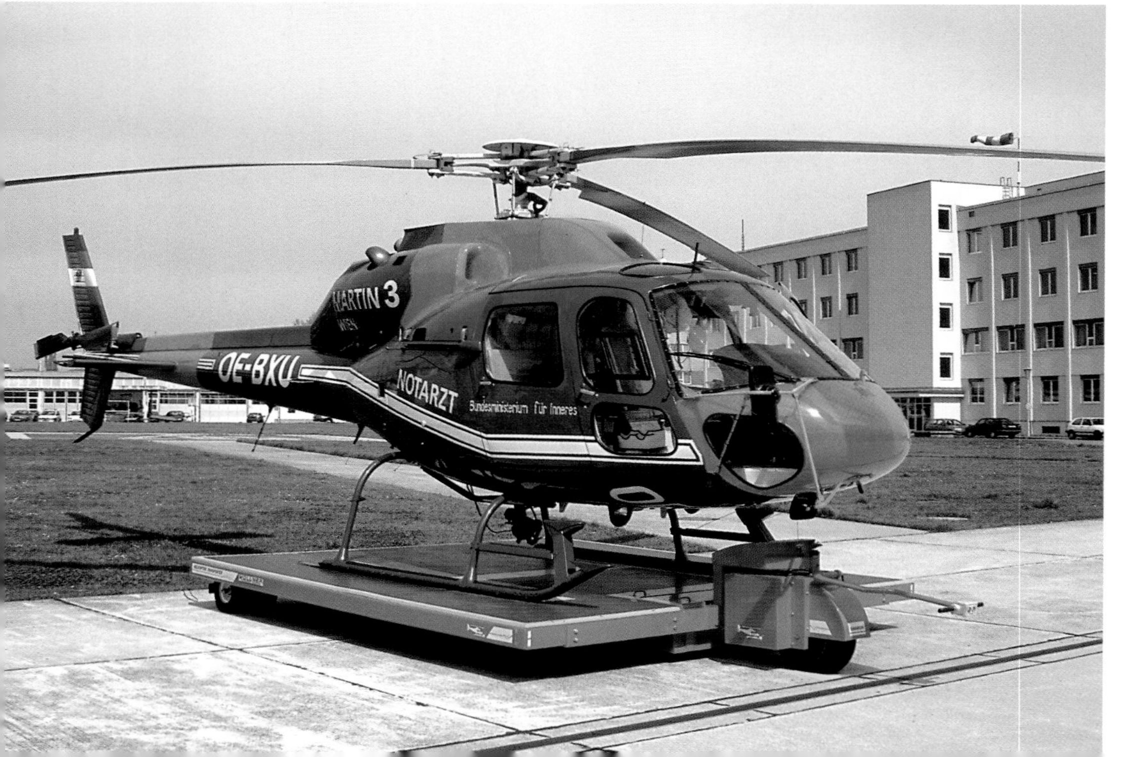

Similarly to the German military, the Austrian Army is available to fight forest fires from the air. Along with several other helicopters, they use the Bell UH-1D seen here.

In forest and surface fires, as well as other medical emergencies, helicopters of the Austrian Ministry of the Interior have been put to use. This Aerospatiale AS-355F-2 Eucreuil (radio code name "Martin 3," registration OE-BXU) is stationed in Vienna.

A helicopter of the Austrian Ministry of the Interior, an Agusta Bell 212. The ministry has about twenty helicopters, which can also be used to fight forest fires.

Aerospatiale SA-315B Lama (# HB-XRD) of the Rhein-Helikopter AG charter company. This firm in Balzers, Liechtenstein, now has an Aerospatiale AS-350 Ecureuil which can be used to fight forest fires.

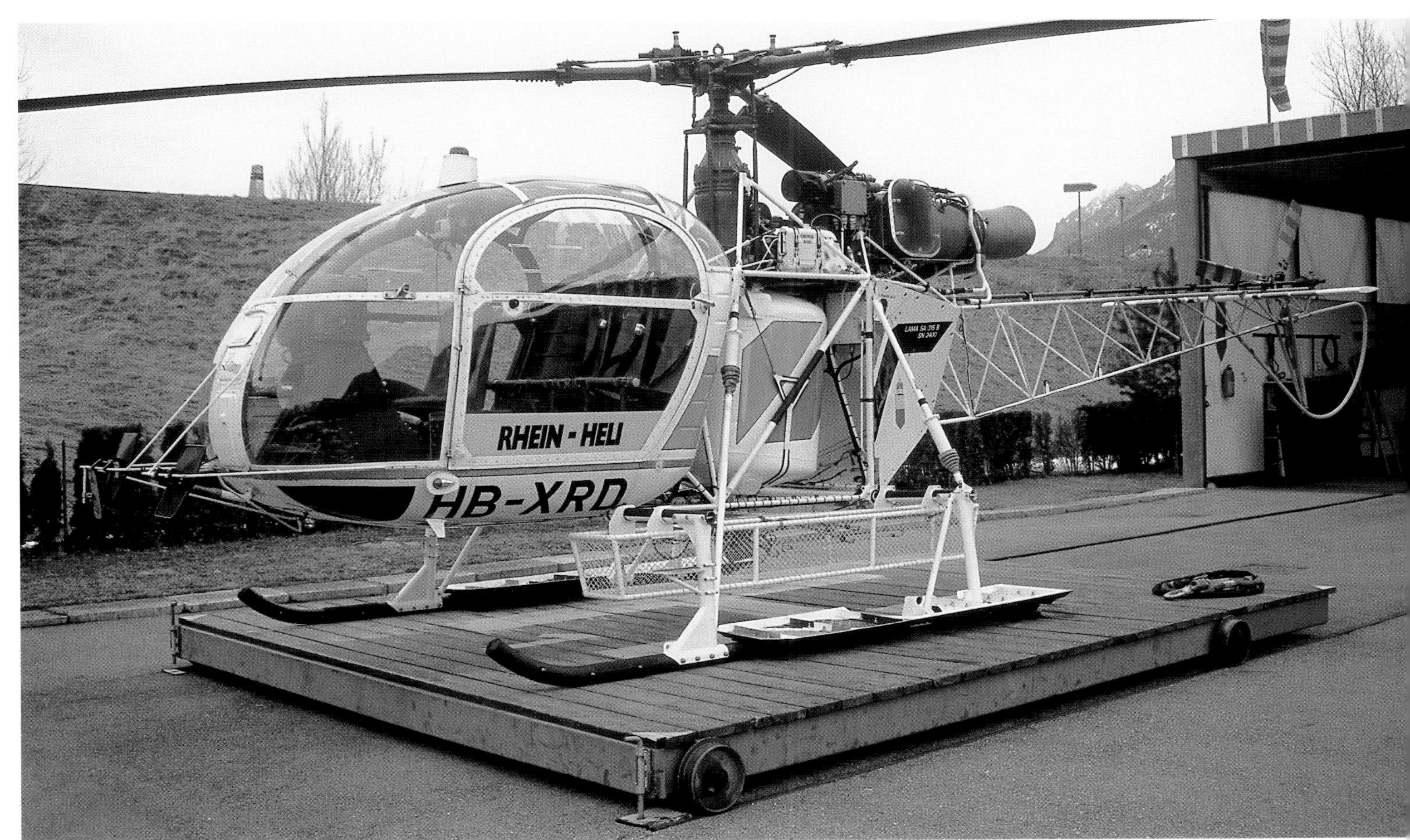

A Bell 206 B JetRanger of the Austrian Ministry of the Interior picks up water to fight a forest fire.

What the ADAC does in Germany, the Austrian Automobile, Motorcycle, and Touring Club (ÖAMTC) does in Austria. This organization also reserves rescue helicopters for medical emergencies, and they can be used if needed to support firefighting action. The picture shows a Eurocopter EC-135T-1 (radio code name Christophorus 6, registration OE-XEG) at the Salzburg airport.

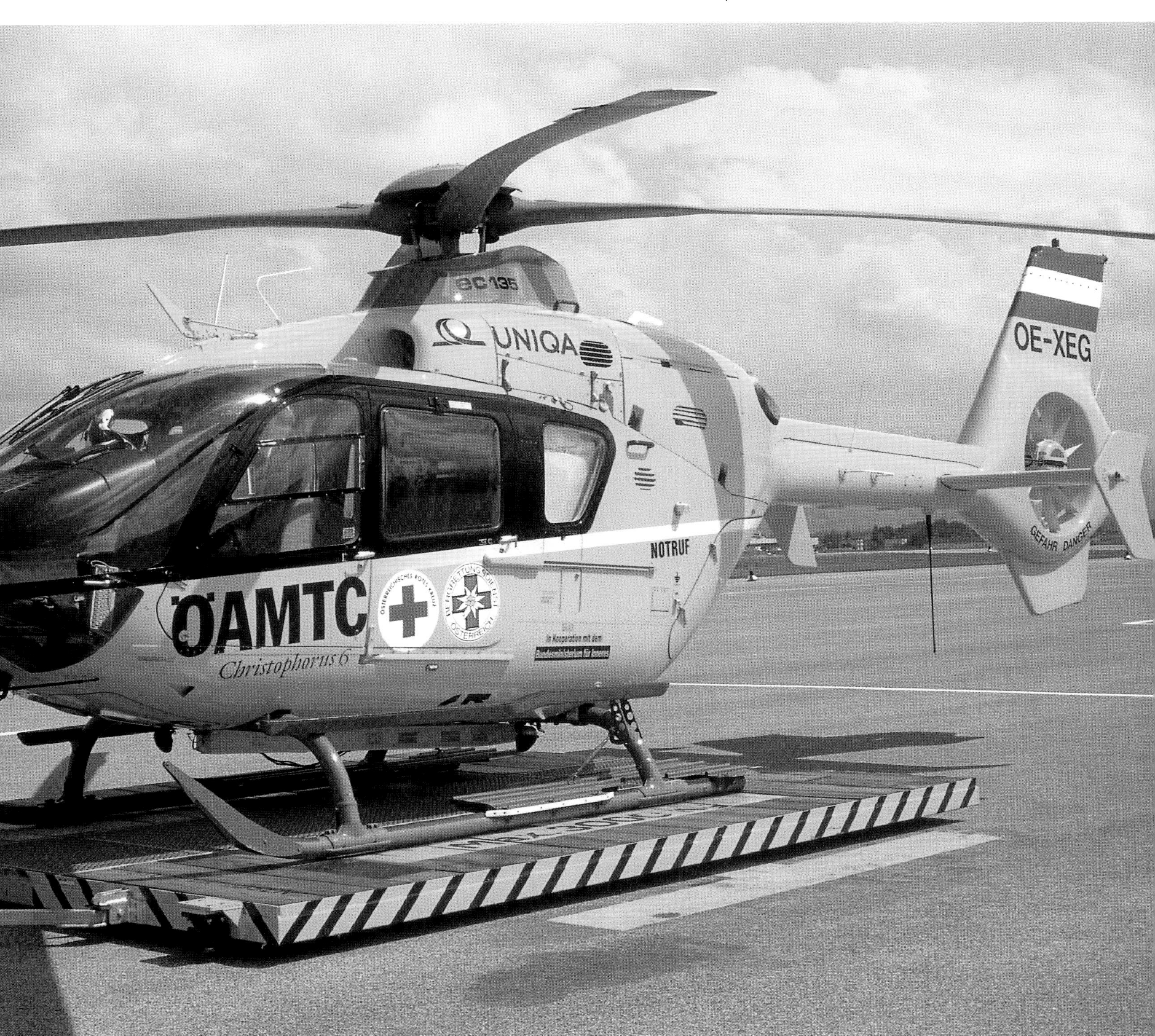

In Switzerland, the Schweizerische Rettungsflugwacht (Rega) is the largest air rescue organization. Along with medical rescue, Rega also makes other assistance flights as well as firefighting runs with Bambi buckets. A special training program trains firemen and others to work with rescue and fire helicopters. The picture shows an Agusta A-109 K2 (# HB-XWL) of the Rega.

The modern aircraft of this organization (including Eurocopter EC-135, AS 350 B3 Eucreuil, SA 315 B Lama, SA 315 B Lama, Agusta A-109 K2) are generally used for forest and surface firefighting as well, and can be equipped with mobile external water containers (Bambi buckets). The Schweizerische Rettungsflugwacht (Rega) in particular trains its personnel and firemen intensively in firefighting with helicopters.

The Swiss Army supplies helicopters on request (such as Alouette III, Aerospatiale AS 332 Super Puma/Cougar) and smaller air-

An Agusta A-109 K2 of the Schweizerische Rettungswacht (Rega) is fighting a surface fire.

planes (Pilatus Porter PC-6) for forest firefighting. In some Swiss cantons, so-called forest fire support points have been set up.

The materials, equipment, and personnel on hand there are also ready for aerial firefighting action. For example, in the canton of Graubuenden there are nine large and nine small forest fire material support points, run in cooperation with the Forestry Department, the Fire Police, and the communities. A forest fire company set up in the fire department numbers some 300 volunteers. The firefighters work closely with civilian protection, forest service, and army personnel.

The Swiss Army also makes helicopters available on request for firefighting work, such as this Eurocopter AS 532 UL Cougar Mark I (military: TH 98, # T-336).

AIR FORCE
-336

European Mediterranean Countries

All European Mediterranean countries (France, Spain, Greece, Italy, Portugal, Croatia) generally have specialized units for forest and surface firefighting from the air. They are equipped not only with airtankers (particularly Canadair CL-215, CL-215T, and CL-415, Dash 8-400, Air Tractor AT-802/F, and C-27J Spartan/MAFFS), but also with fire helicopters and helitankers.

See also:
Waterbombers (Scoopers)—Amphibian Planes.

France

Technically and tactically, France depends mainly on the concept of amphibian fire aircraft (scoopers) made by the Bombardier firm of Canada. These are the only planes in the world that are produced especially as airtankers.

The French Securite Civile (civil and catastrophe service) presently owns eleven Canadair CL-415. There are also ten S-2T Tracker Turbo FireCats and two Dash 8-400 on hand for fighting forest and surface fires. Additional aircraft are to be purchased. The central fire aircraft base of the Securite Civile is located at the Marseille-Provence airport at Marignane. Amphibian planes of the Consolidated PBY-5A Catalina type were formerly used.

Helicopters from numerous charter companies (contractors) are also integrated into firefighting, for example, Eurocopter AS-350B Eucreuil and Bell B-205A (Helipaca).

France has an extensive fleet of firefighting airplanes. Besides the well-known Canadair amphibian airtankers, these Grumman TS-2A Firecats from Conair serve the French Securite Civile in fighting forest and surface fires. The picture shows Tanker 07 (# F-ZBEY) at the fire airbase of Marignane near Marseilles.

Several Fokker F-27 Friendship (in the picture, Pelican 72) also belong to the firefighting aircraft fleet of the Securite Civile. They are former turboprop commercial airliners that were converted into firefighting planes.

Spain

Since 1971, Spain has used mainly Canadair amphibian planes to fight forest and surface fires. The Spanish Air Force (Fuerzas Aereas) and the national forest service have, in all, 15 Canadair CL-215T and seven Canadair CL-215.

The planes are stationed at, among others, the Aerea Getafe military base in Barajas and the Torreon base near Madrid. During the forest fire season in the hot summer months, three military airfields become temporary fire airbases: Santiago de Compostella in Galicia, Reus in Catalonia, and Jerez de la Frontera in Andalusia. As a rule, between June and September a Canadair CL-215T—besides several Air Tractor AT-802—are stationed at the Pollensa base on the north side of the island of Mallorca. They are responsible for the Balearic Isles and the southern part of mainland Spain.

A Canadair CL-215 is also operated from the Girona base by the Cegisa.

In addition, some 30 Air Tractor AT-802 and two Air Tractor AT-802F Fire Boss (Avialsa) and several PZL M-18 Dromader of private owners are available for firefighting.

A French Convair 580 firefighting airplane (Tanker 52, # C-FKFA) of the Canadian Conair Aviation charter company. The plane is stationed at the Marignane fire airbase near Marseilles.

Private charter companies (contractors) also run a large number of helicopters that are used to fight forest and surface fires, including Mil Mi-8 (Heli Air Services), Kamov KA-32 (PANH-Teneriffa), and Bell B-212 (TAF Helicopters).

The Generality of Catalunya (the Catalan district government) used fire helicopters and helitankers on its own, including MBB-Kawasaki BK-117, Bell B-212, and MBB Bo-105 CBS.

Greece

In Greece too, amphibian planes are at the head of the aircraft fleets for fighting forest and surface fires. The Greek Air Force now has 15 Canadair CL-415 and 14 Canadair CL-215 firefighting planes.

In addition, the Greek government has had eight C-27J Spartan military transport planes (similar to the Lockheed C-130) equipped with the American modular Airborne Fire Fighting System (MAFFS).

As in other Mediterranean countries, PZL M-18 Dromader Single Engine Air Tankers (SEAT) are also used.

The Greek Mediterranean islands are also protected by firefighting planes. The Greek Civil Guard (Politki Prostasia) stations two Canadair firefighting planes on Samos during the summer months. Forest firefighting is carried on there along with the fire department, forest fire service (some 40 fire planes in all), and the army.

Italy

Italy has a very extensive array of aircraft for forest and surface firefighting. At the bases of the Canadair agent SOREM in Genoa, Treviso, Rome, Olbia, Trapani, and the Calabria region there are 14 state-owned Canadair CL-415 plus two Canadair CL-215 amphibian firefighting planes of the Protezione Civile for firefighting.

The Italian national forest service, Corpo Forestale dello Stato (CFS) has a large number of special helicopters ready for forest firefighting, including the Agusta Bell 412 CS, Agusta Bell 412 EP, Bredanardi-Hughes NH-500 D, Bredanardi 369 HS, Bredanardi-Hughes NH-369 D, Air Tractor AT-802, Sikorsky S-64F Erickson Skycrane, and one Piaggio P-180 Avanti airplane for personnel and material transport.

The French firm of Air Attack Technologies (AAT) of Marignane near Marseilles has a Cessna 208B Caravan (# D-FLIP) as an infrared airplane. The plane is equipped with extensive camera and documentation technology and is used to observe large forest and surface fires.

During the forest fire season, this Eurocopter AS 350 B2 (# F-GJCM) is stationed on the French Mediterranean island of Corsica. The helitanker is equipped with a Simplex Fire Attack tank system and chartered from Corseus Helicopteres.

Also stationed temporarily on Corsica is this helitanker of the same type. The water tank on the bottom can be seen clearly in the picture.

Italian fire departments also use helicopters (Elicottero) for firefighting. They are stationed at twelve bases in Arezzo, Bari, Bologna, Centro Aviazione, Catania, Genoa, Pescara, Salerno, Sassari, Torino, Varese, and Venice and are used, along with search and rescue missions, mainly for fighting fires. Such helicopters as the Agusta-Bell AB-204, AB-206, AB-412, and AB-412EP, Aerospatiale AS-365 N3 Dauphin (Trento region), and Agusta A-109E are used.

Along with this extensive array of aircraft used to fight forest and surface fires, other helicopters and Single Engine Air Tankers (SEAT) belonging to private businesses, rescue organizations (Soccorso Sanitario, White Cross, Red Cross, Guardia Costiera, the Polizia di Stato, Carabinieri, Guardia di Finanza, and mountain rescue organizations (including Aiut Alpin Dolomites) are available in all regions of the country. The aircraft used there are generally the same models that are listed above.

Portugal

For forest and surface firefighting there are two state-owned Canadair CL-215 (SOREM) planes on hand in Portugal, plus a few PZL M-18 Dromader. The Portuguese charter companies Heliportugal (Kamov K-32 etc.), Helisul, Helibravo, and the Polish Panstvove Zaklady Lotnicze (PZL) were formerly involved in firefighting.

In Spain the Cegisa (CEG, Gestair Group) uses Canadair CL-251 for forest and surface firefighting. This firm from Madrid (the Torrejon de Ardoz fire airbase) has six more of these planes on hand.

RESCUE
1269

The Grumman G 164 Ag-Cat Single Engine Air Tanker (SEAT) is seen at the Tatoi military airbase in Greece. The plane carries 1890 liters of water for firefighting.

The French Securite Civile has two Dash 8 (Bombardier/de Havilland DHC-8-402Q) planes, one used for firefighting, the other as a personnel transport.
The picture shows the Pelikan 73 firefighting plane (# F-ZBMC) taking on fire retardant at the Marignane air base near Marseilles, France.

For large forest fires, the Portuguese Ministry of the Interior charters firefighting planes and helicopters from international firms (including more PZL M-18 Dromader, Canadair CL-415, Russian Beriyev Be-200 or American S-64 Skycranes) or requests air support in the framework of national neighborhood help (EU, UN, Swiss Army: three AS 332 Super Puma, in August 2004).

In June and July 2006, a Russian Beriyev Be-200 amphibian firefighting plane was in service in Portugal for two months (from the Monte Real air base support point)—which cost the Portuguese government some 1234 million Euro.

For the year or the forest fire season of 2008, Portugal plans to obtain four large airtankers and six medium and four small fire helicopters to fight their forest and surface fires, which are becoming more frequent and extensive.

Croatia

Some eighty members of the Zagreb professional firefighters form a forest fire unit that is active not only in Croatia itself, but also, when requested, in nearby countries within the neighborhood assistance program. During the summer forest fire season on the Mediterranean coast, some of their forces are assigned to support points in Dubrovnik, Zadar, and Sibenik.

The unit practices close cooperation with the air support units in the country.

Eurocopter AS-332L1 Super Puma (# SX-HFF), a fire and rescue helicopter of the Greek firefighters, is seen at Athens International Airport. The copter can carry up to 3500 liters of water in Bambi buckets.

In state service there are now four Canadair CL-415 amphibian firefighting planes in Croatia, plus three Canadair CL-215. In addition, various fire helicopters of commercial charter contractors are available (such as Eurocopter BK-117 and Eurocopter EC-145).

In serious emergency cases, international requests for further air support can be approved by the Ministry of the Interior. In forest and surface fires in Croatia, Beriyev Be-200, Douglas DC-10, and Ilyushin Il-76 firefighting planes, among others, have already seen service.

Czech Republic

Aerial firefighting in the Czech Republic is within the jurisdiction of the Czech Ministry of Agriculture. The work and action, though, are coordinated by the Ministry of the Interior, which also supervises the Czech professional firemen. The guidelines for firefighting and control flights over Czech territory (not including military zones and national parks) were set up in 2005. The services of private charter companies have been established since 1933 and agreed on by contract. Helicopters of the Czech police are also included in Ministry of Agriculture contracts and fly control missions for the forest fire patrol as well as fighting fires.

The Greek Ministry of the Interior also uses large American-made Erickson Air-Crane helicopters from Oregon to fight forest and surface fires. Here a Sikorsky S-64E Skycrane (Helitanker 735, # N173AC) is seen at the Eleusina military airbase in Greece. The helicopter's tank holds up to 9000 liters of water.

The Czech Republic is divided into 14 districts, and ranks them at risk levels A, B, and C in terms of forest and surface fire danger. For A sectors there is always a forest fire airbase with a patrol plane and a firefighting plane on hand, which must be in use between 10:00 AM and 6:00 PM. For B and C sectors, there must be at least one patrol plane on hand at the bases. The flight schedules are made in advance, up to 24 hours ahead in B sectors, up to 40 hours in C sectors.

Problems for the organizations and businesses involved in fighting forest fires are caused by contracts that last only one year. This short planning time hinders firms' investments in new equipment, among other things. At present, former agricultural planes are used.

In 2006 there were nine forest fire airbases in the country, in Tachov, Hosin, Holesov, Mnichovo Hradiste, Jihlava, Zamberk, Plasy, Zabreh, and Znojmo. Aircraft that saw service included fifteen Antonov AN-2 Andula, Z-37T Turbo Cmelak, PZL M-18 Dromader airplanes and the Bell B-412 helicopters of the Czech Police.

Poland

The following firefighting airplanes and helicopters are known to be used in Poland,

among others: PZL-Mielec M-18 Dromader (such as Zak ad Us ug Agrolotnicych.ZUA) Antonov An-2, Mil Mi-2 (Heliseco), PZL-Svidnik Mi-2, and PZL W3 Sokol (Heliseco).

The provision and use of firefighting planes (mostly SEAT) are assigned by, among others, the state forest and forest-economy agency (Panstovove Gospodarstvo Lesne/Lasy Panstvove) to private charter companies.

Slovakia

As for the concept of air support in forest and surface firefighting in Slovakia, we can look at the information on the Czech Republic. When the two states were still united as Czechoslovakia, the concept was worked out jointly in 1992.

At present—along with some Single Engine Air Tankers (SEAT) owned by civilian charter companies—one fire helicopter each of the Ministry of the Interior (Mi-171, Bratislava airport), the Slovakian Army (Mi-17, Presov military airbase), the Air Transit Europe firm (Mi-8, Poprad airport), and Aero Slovakia (Zlin Z-137, Kosice airport) are on hand for forest and surface firefighting.

For patrol flights, there are also small airplanes of Aero Slovakia in use: Cessna C-150, C-152 and C-172, plus Zlin Z-42 and Z-37.

MBB-Kawasaki BK-117 C-1 fire helicopter (# SX-HFH) of the Greek fire department at Athens Airport.

This Russian Mil Mi-26T fire helicopter (# RA-06295) of the Bulgarian Scorpion air charter company is stationed at Thessaloniki, Greece.

A fire helicopter of the Italian state forestry agency, Corpo Forestale dello Stato (CFS), an Agusta AB-412EP (# I-CFAF). The agency is in charge of numerous special fire and observation helicopters.

Just as exemplary is the equipping of Italian fire departments with helicopters for firefighting and rescue tasks. The picture shows an Agusta-Bell AB-412 (Vigli del Fuoco/VF 55, # I-VFOG) in action.

Above: The most modern fire helicopter of the Italian Vigili del Fuoco (firefighters) is the Agusta A-109E (VF 80).

Lower left and right: Along with several firefighting airplanes and helicopters of the forest and forest fire organizations, fire departments, military and civil guard, helicopters of the rescue organizations can also be called on for firefighting. The pictures show two interesting Aerospatiale AS-319 Alouette III of the White Cross in Bozen, South Tirol.

A mountain rescue helicopter of the Aiut Alpine Dolomites from South Tirol, Italy. The Eurocopter AS-350 B-3 Ecureuil (# I-AMVG) was replaced by one of the same type (# I-AMVM) and then by a Eurocopter EC-135 T-2 (# D-HDOL, now I-HALP).

The Aiut helicopters can be equipped with a Bambi bucket for fighting forest and surface fires. The picture was taken at the Sanon Hut, the copter, then I=AMVG, is in the background.

A Canadair CL-415 amphibian airplane is seen fighting a fire in the coastal region of Croatia.

A Czech Single Seat Air Tanker (SEAT), a former Zlin 137T Agro Turbo agricultural plane (# OK-EJA).

A Czech agricultural plane is refueled at an airport in Slusovice.

Russian Antonov AN-2R biplanes are also used for forest and surface firefighting in the Czech Republic.

This Piper PA-28-181 Archer (# OK-FLY) of the Czech Fly For Fun charter company, located not far from Prague, can be used for observation and oversight flights of the state forest and surface firefighting agency.

A Bell 427 (# OK-EMI) of Alfa Helicopter, from Olomouc, Czech Republic. Rescue helicopters of this type can also be called in for firefighting, if needed.

helicopter

ec135
eurocopter
AIR RESCUE SERVICE
LETECKÁ ZÁCHRANNÁ SLUŽBA

A Bell 412 HP helicopter of the Czech Police from Prague. Police helicopters are also used in the Czech Republic for forest and surface firefighting, as well as for fire observation.

This Eurocopter EC-135 T-1 (# OK-DAS) of Delta System Air is in use by the professional fire department in Ostrava, Czech Republic. The private charter company, which has numerous other helicopters and airplanes (including Mil Mi-2, Aerospatiale AS-355F-2 Ecureuil), is integrated into the rescue service (Letecka Zachranna Sluzba). Their helicopters are also available for firefighting.

A Czech Zlin Z-37A (# OK-AGR) of Agro Air in Zamberk.

A look at Poland: A PZL-Mielec M-18 Dromader is being filled with water from tank trucks (TLF 16/25 Star 244) of the Koscielek and Rudniki fire departments at the airfield of the Aeroclub Czestochova. The plane belongs to the service agency for agricultural flights (Zak ad Us ug Agrolotniczych/ZUA) in Mielec, a branch firm of the EADS PZL Warszawa-Ok'cie S.A.

A Mil Mi-2 fire helicopter (# SP-SLM) of the Polish Heliseco in action at a forest fire near Ciarka/Opole. Heliseco is Poland's largest helicopter charter firm, owning 22 PZL W3 Sokol and 48 PZL Mi-2 copters (below). About 60% of them are used in forest and surface firefighting in Poland, and also in Spain and Portugal since 1988.

Russia has the world's second largest firefighting aircraft, after the USA. One of them is the Beriyev Be-12P 200 Chaika (# RA-00041) amphibian firefighting plane of the Russian Forest Fire Guard, seen here dropping water. The plane's water tank holds 9000 liters.

Russia and ex-Soviet States

Russia has not only the largest contiguous woodlands in the world, but also the most forest fires, with immense damage to landscape and vegetation. Every year, statistics show, as much woodland is destroyed as its wood is needed for consumption by the former Soviet Union. Statistics register between 20,000 and 30,000 forest and surface fires a year in all of Russia.

So it is no surprise that Russia traditionally has extensive and capable technical and tactical resources for fighting forest fires from the air. The Russian Ministry for Civil Defense, Emergency and Elimination of Consequences of Natural Disaster (EMERCOM) consistently supports the use of large airtankers for fighting forest and surface fires. The Russian aerial firefighters have the Ilyushin IL-76 TP, presently the world's largest and most capable airtanker, at their service.

Since the mid-thirties the government aerial firefighting agency, Avialesookhrana, of the Russian National Forest Fire Center, with headquarters in Pushkino and Moscow, has provided preventive and aggressive forest fire protection for all of the Soviet states.

From 24 regional aerial fire centers and fire airbases, over 800 million hectares of woodland are observed and protected. The Avialesookhrana works either commercially for the regional forest and woodworking industry or under contract to the state to protect state-owned land.

The Russian state forestry and firefighting agency, Avialesookhrana, uses Beriyev Be-12P-200 Chaika planes like this one (# RA-00046).

The organization, with some 8000 men—including 600 regularly employed pilots and 4500 smokejumpers—has many aircraft of its own, over 300 of the organization itself plus another 300 for every forest fire season, chartered chiefly from Aeroflot, the Russian airline, including numerous light (ICS Type III) and heavy (ICS Types I and II) airtankers and fire helicopters. Russia's aerial firefighting system also includes, as in the USA, smokejumpers with AN-2 jump planes and heli-rappelling with MI-8 helitack copters.

Preventive fire protection in Russia's gigantic forests is administered not only through ground-based surveillance by various established and equipped forest fire patrols (sizes 1 to 3), and by an air-supported system of patrol flights. Thus some 85% of all forest fires can be discovered and fought early.

Problems in the present concepts of preventive and aggressive forest fire protection include the frequent lack of coordination between ground and air units, insufficient financing of equipment, materials, and personnel, and an incomplete satellite surveillance system.

The best-known aircraft used to date in the Russian states to fight forest and surface fires are the Beriyev Be-12 Chaika/Seagull amphibian, Ilyushin Il-76 firefighting plane, Beriyev Be-200, A-40 Albatross and A-200 Albatross

Australia: One of the many helicopters—Helitanker 402, a Bell 214B (# N49732) of McDermott Aviation—chartered during the fire season, takes on water in the Blue Mountains west of Sydney.

amphibian planes, Mi-8 fire helicopter-helitack, Mi 8MTV fire helicopter-jump plane, Kamov 32 (fire helicopter-helitack, Antonov AN-2P fire and jump plane, and Antonov AN-28 firefighting plane.

Australia

The extensive regions of Australia are overseen by several forestry and firefighting organizations: the Forest and Wildland Firefighting agencies of the states, the national Australian Fire Authorities Council, and the National Aerial Firefighting Centre:

- South Australia: South Australian Metropolitan Fire Service, South Australian County Fire Service, Woods and Forests Department, National Parks and Wildlife Service.
- Tasmania: Tasmanian Fire Service.
- Victoria: County Fire Authority, Department of Conservation and Environment.
- Capitol District: A.C.T. Bush Fire Council.
- New South Wales: New South Wales Rural Fire Service.
- Queensland: Queensland Fire Service.
- Western Australia: Bush Fire Board of Western Australia.
- New Zealand: New Zealand Fire Service.

All the organizations, because of the vegetation and weather conditions in a great part of the continent, have emphasized the fighting of forest, bush, and surface fires, and thus have their own aerial firefighting agencies.

In Australian aerial firefighting, a great number of different aircraft are in use—mostly helicopters and Single Engine Air Tankers

(SEAT). The New South Wales Fire Service alone has over 100 patrol and firefighting airplanes. A great many of the airplanes and helicopters used for firefighting, though, are provided through contracts with charter contractors.

To fight forest, bush, and surface fires in Australia, the following aircraft types, among others, are used:

Turbine Thrush agricultural planes, Air Tractor AT-502 and AT-802, PZL M-18 Dromader, Bell B-206 JetRanger, B-206L LongRanger and B-407, Hughes H-500D, Aerospatiale AS-350 Squirrel, Agusta A-109, small (SEAT) planes, and fire helicopters.

Kawasaki BK-117, Bell B-204 (UH-1H), B-205, B-212, and B-412 medium fire helicopters and helitankers.

Skycrane CH-54 B (7500-9000-liter tanks), Kaman K-Max, UH-60 Blackhawk, SH-60B Seahawk, Kamov 32, Bell B-214, Sikorsky S-64 Skycrane heavy fire copters and helitankers.

Piper Warrior, Cessna C-182, C-210, C-310, C-337, and C-404 Titan light patrol airplanes.

Africa

The African continent is far away, and it is surely not easy to obtain reliable information from there on the fighting of forest, bush, and surface fires. During my research, though, it became clear that in some countries and regions of Africa—notably in South Africa—firefighting from the air is carried on by firefighting and fire-protection organizations, forestry agencies, and private charter companies.

For example, in the Sub-Sahara region there is a Wildland Fire Network (AfriFireNet) that includes some 800 firefighters. Specially trained by the Wildland Fire Training Center Africa (WFTCA), 14 "hotshot crews" of 2 men each are trained by American examples. The forest fire units are set up as 40 fire services. For aerial firefighting, six helicopters (Mi-8 with 3500-liter water tanks, Kamov 32 with 4000-liter tanks) and some 25 airplanes, usually smaller types (Turbo Thrushes with 1500- and 2000-liter tanks, PZL M-18 Dromader)

Zululand, Africa: Mil Mi-8MTV-1 (# ZS-RIV) of the Zululand Fire Protection Service, takes up water from an open source.

are used, including 15 aircraft for surveillance of the extensive areas.

In Zululand (southern Africa) they began at the beginning of the 1980s to use agricultural airplanes to fight forest and surface fires. The leader in aerial firefighting was the commercial firm of Orsmind Aerial Spray Ltd. (Orsmind Aviation) from Bethlehem. At first one Piper PA-25 Pawnee was used; today there are some 30 aircraft, including Single Engine Air Tankers (SEAT), such as the Ayres S2R-T34 Thrush, in use.

In addition, helicopters, including the Bell B-407 and B-205, Kamov 32, and Mi-8 are used in Zululand, as well as Cessna C-182 patrol planes.

In the province of East Transvaal (South Africa), the private Forest Fire Association (FFA) is at work. Their main base is in the city of Nelspriut, with regional bases in Peat Retief, Warburton, and Tzaneen. The organization has seven patrol planes (Cessna C-182 and C-206), twelve PZL M-18 Dromader and Ayres S2R-T34 Thrush Single Engine Air Tankers (SEAT), and one Russian Mi-8 helicopter with 3500-liter Bambi bucket as fire and helitack copters.

South America

Argentina

Since the mid-fifties, experience with American aerial firefighting has been gained in Argentina. As in the USA, forest and surface firefighting from the air developed out of agricultural aviation. At first, former agricultural planes were fitted with tanks and spraying devices, later to be replaced by Single Engine Air Tankers (SEAT). Today both state forestry and forest fire organizations and private charter companies (including the Agencia Cordoba Ambietenes and the Direccion General de Aeronautica de Cordoba) carry out forest firefighting from the air.

Along with the preferred SEATs (such as the Air Tractor AT-502), fire helicopters (such as the Agusta A-109, using Bambi buckets) and small and medium airplanes (such as the Cessna 206 patrol and lead planes, Cessna 188 and Cessna 350 Air King) are used for observation and patrol flights.

Chile and Cuba

In both countries, the Polish PZL M-18 Dromader, among other planes, are used for firefighting.

Asian Countries

In most Asian countries, large airtankers and helitankers are available for firefighting. Aircraft of the professional firefighters and the military are generally used.

Firefighting helicopters generally fulfill multiple functions. Along with firefighting (buildings, forests, and surfaces), search and rescue missions, water rescues, and medical rescues are among their tasks.

China

The following firefighting planes are known to be used in China:
PZL M-18 Dromader, Harbin SH-5 airtanker (Harbin Aircraft Manufacturing Company, 8000 liters), and Harbin Z-5 helicopter.

Thailand

Forest and surface firefighting in Thailand has traditionally not had a high priority. More and more, though, they are becoming aware of the problem of nature and landscape damage. Thus the Thai Army now carries out increasing patrol and observation flights along with the forestry authorities.

Two Canadair CL-215 are stationed with the Royal Thai Navy. These planes are used mainly as observation craft.

Japan

The fire protection and firefighting system in Japan is highly developed and stands out with its modern equipment, aircraft, and training concepts. Numerous professional fire departments in Japan have helicopters, which are used mainly as medical rescue craft, for rescuing people from skyscrapers and mountains, or for logistical support of ground-based firefighters. In addition, most Japanese fire helicopters are available for forest and surface firefighting.

For example, the Tokyo Fire Department has four Eurocopter AS-365 Dauphin II helicopters. equipped with external Bambi buckets, ready to fight fires. Two more helicopters, Aerospatiale AS-332 Super Pumas, have firmly attached water tanks (helitankers) with a capacity of 1200 liters. In addition, each of the two helicopters can carry a 4500-liter Bambi bucket.

Control, monitoring, and transport aircraft of the South Korean Forestry Administration of type Let L-410 UVP-E 9 Turbolet (# FP 502).

The Fukuoka Fire Department (south of Shinkansen) has two modern Agusta A-109 multipurpose helicopters. Both are used for the tasks noted above, as well as for firefighting.

Eurocopter-Kawasaki BK-117 helicopters are used by, for example, the professional fire departments of Kitakata, Setouchi, and Osaka. They are used primarily for rescue work and firefighting, in highlands (helitack) and to fight forest and surface fires with Bambi buckets. The Japanese PS-1 fire airplane is an amphibian airtanker built by the Japanese firm of Shin Meiwa.

The Antonov An-26 (# RA-26002), with its 4000-liter water tank, ranks among the medium-sized firefighting airplanes of the Aviale-sookhrana. These planes are also used to transport smokejumpers.

The Beriyev Be-200Chs (# RA-21515) of the Russian Catastrophe Protection Ministry is the world's largest amphibian firefighting airplane. Its water tanks can carry up to 12,000 liters of water, which can be picked up in a few minutes while gliding.

AH-2

Among the smaller firefighting planes of the Russian Avialesookhrana is this Antonov An-2 (# RA-56485). The An-2L and An-2VA versions have been equipped for firefighting.

The Russian Kamov TA 32 A1 fire helicopter—note the two main rotors, one above the other, with a diameter of 15.90 meters each—are used to fight forest and surface fires (with 5000-liter Bambi buckets) as well as for rescues from tall buildings, search missions, and medical emergencies. The helicopters carry two pilots and two firefighters.

Left: Mil Mi-8 MTV-1 (# 102) of the Lithuanian Air Force.

Above: This Mil Mi-14 BT (# 59-TAJ) is operated by Aerotec International and was formerly used by the National People's Army of the ex-DDR.

A Russian Mil M 26 T firefighting helicopter of the Catastrophe Protection Ministry (Emercom).

A Chinese Aerospatiale AS-332L Super Puma fire helicopter (# B-7959) of the City Offshore Helicopter Co. Ltd. (COHC), China's largest helicopter charter company.

Just below: Four Aerospatiale SA-365 C Dauphin II helitankers (This is # JA 9569) belong to the Tokyo professional fire department in Japan. They are used regularly for air support in major emergencies (including searches and rescues, in medical emergencies, and as command planes, as well as for firefighting with a 1200-liter fixed tank.

Bottom: This Russian Mil Mi-171 Helion Procopter is used by the Malaysian firefighters to fight fires in Malaysia. It was rebuilt by the Airod Snd. Bhd. firm in Subang, Malaysia. For firefighting action, Bambi buckets with 1500- and 3000-liter capacity are used. The copter is also available for search-and-rescue missions.

Kamov KA-32T, formerly battle helicopters of the South Korean Army, were rebuilt as fire helicopters (helitankers) and made available to the state forest administration. Especially impressive features of the Kamov KA-32T are the two rotors, one above the other. An interesting detail: In South Korea too, helitankers are equipped with the American Simplex Fire Attack tank system, which is used worldwide.

The South Korean Forest Administration also uses Kamov KA-32T helitankers for helitack crews. An interesting detail: The copters are fitted with two water snorkels to pick up water.

This Ansat fire helicopter (# FP 301) of the South Korean Forest Administration was made by the Kazan Helicopter Plant is Russia. The agency, which is subordinate to the Ministry of Agriculture, has its own forest fire department (Forest Protection Bureau, Forest Emergency Situation Team).

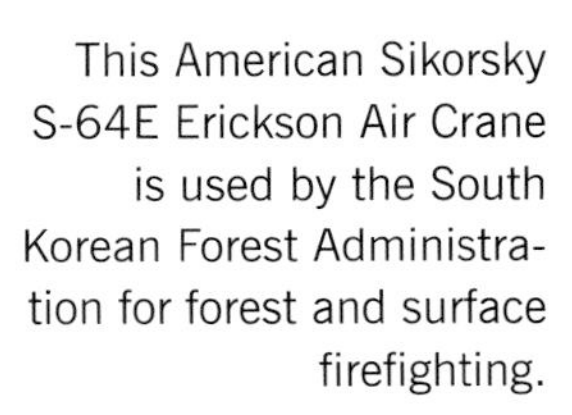

This American Sikorsky S-64E Erickson Air Crane is used by the South Korean Forest Administration for forest and surface firefighting.

Waterbombers—Canadair Amphibian Planes

The amphibian firefighting airplanes made by the Canadian firm of Bombardier Aerospace in Montreal are very definitely special! They are the world's only airtankers built especially for firefighting. They are particularly widespread in the Mediterranean countries of Europe.

The Canadair CL-215 amphibian airtanker was the first answer to the request of Canadian forest and fire managers for a new, more effective, and powerful airplane for fighting forest and surface fires.

This first amphibian airtanker of a completely new generation was fitted with two water tanks holding a total of 1410 gallons (2673 liters each, total 5346 liters), two foam tanks (800 gallons/300 liters each), and two Pratt & Whitney R 2800 double-star motors of 2100 HP (1565 kW) each. The top speed of the new airtanker was 304 kph.

The first plane was delivered to the French Protection Civile in June 1969, and by 1989 a total of 125 planes from five production runs were sold, primarily to Europe.

In addition to their use in firefighting, Canadairs were also used as search-and-rescue planes, military and coast guard observation craft, and as civilian passenger and freight planes.

The Canadair CL-215 was, and still is, used in Spain (30 of them), Greece (16), France (15), Italy (5), Yugoslavia (5), Thailand (2), Venezuela (2), and Canada itself (49).

A logical further development was the CL-215T. As a hybrid aircraft with the upgraded performance of the successor type and its predecessor's tank capacity, it forms a kind of technical transitional stage between the Canadair CL-215 and the Canadair CL-415, which came on the market in 1994. The first plane of this type was sold to Spain in June 1991.

The technically further-developed Canadair CL-215T is also equipped with two water tanks holding a total of 1410 gallons (5346 liters), two foam tanks (80 gallons/300 liters), but uses two Pratt & Whitney 123AF turboprop motors (2100 HP/1565 kW). Its top speed is now 365 kph.

At this time the Canadair CL-215T are still used in Spain (15) and Quebec (2).

The most modern and up-to-date version of the Canadian amphibian airplanes is the Canadair CL-415. It has been used since 1994 and is equipped with four water tanks holding a total of 1621 gallons (6137 liters), with foam mixing (one or two foam tanks), 80 gallons (300 liters) each, and two Pratt & Whitney 123AF turboprop engines, each producing 2380 HP (1775 kW). The top speed of the CL-415, like that of its predecessor, is 365 kph.

While the Canadair CL-415 has been used primarily in Europe, the use of amphibian airtankers in American aerial firefighting is very restrained. Only the US states of North Carolina and Minnesota have bought a total

A Green amphibian firefighting plane, Canadair CL-215 (# 1045) at the Eleusina military airfield.

of three Canadair CL-415. In 2006 the Los Angeles County Fire Department (LACoFD) put two Canadair CL-415 planes into service provisionally. For tactical as well as financial reasons, no other amphibian airtankers will probably be used there to fight forest and surface fires.

In all, 53 Canadair CL-415 airtankers have been sold to the following countries as of 2001: France (11), Greece (8), Italy (14), Croatia (3), and Canada (17).

Canadair planes, in their characteristic all-steel flying boat style, are not only amphibian, as are many other amphibian planes that can take off from and land on open waters, but they can also pick up water into their tanks while flying at speeds up to 130 kph! To do this, a scooper of about 40 cm is extended behind their tail angle to pick up the water in low flight and fill the Canadair's tanks.

Via two (CL 215, 215T) or four (CL-415) hatches, the water can be dropped in doses over the scene of the action. Foam from installed tanks can also be added to the water.

The Canadair CL-415 needs a distance of some 665 meters to land on the water, some 800 meters to take off, and a surface of 1341 meters to pick up water and then take off. The water tanks can be filled within twelve seconds.

This Canadair CL-215 (# 43-21 1116) of the Spanish Air Force (Ejercito del Aire) is "Ready for takeoff."

Airtanker 262 (# N262NR), a Canadair CL-215, is stationed in Alaska during the forest fire season.

Bottom: A Canadair CL-415 (Tanker 48, #F-ZBMG) of the French Securite Civile.

The Canadair amphibian firefighting planes are particularly widespread is southern Europe. Here is a Canadair CL-415 (No. 877) of the Croatian Air Force. Croatia presently uses four of these planes (No. 844, 855, 866, 877) plus three Canadair CL-215.

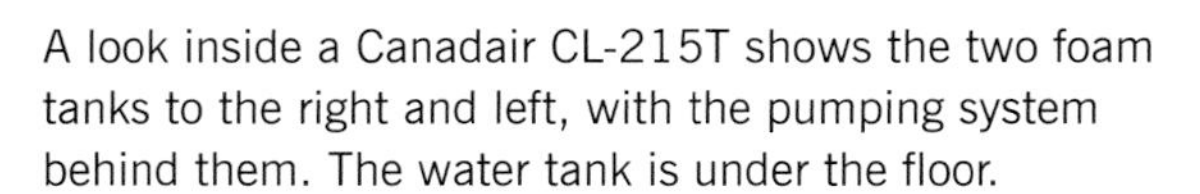
A look inside a Canadair CL-215T shows the two foam tanks to the right and left, with the pumping system behind them. The water tank is under the floor.

An interesting picture—Fire Plane 44 (# F-ZBME) of the French Securite Civile in gray! Later the planes were repainted in the organization's standard colors (see picture at right).

Below: A Greek Canadair CL-415 (No. 2055) at the Eleusina military airbase.

A Canadair of the French Securite Civile (Pelican 42, F-ZBEU) plows along steadily and picks up water from the wavy sea.

Bottom: The Air Tractor AT-802F Fire Boss also ranks among the amphibian firefighting planes that serve successfully in the USA, Canada, and Europe.

Along with the Canadair, the Canadian-American consolidated PBY-6A Catalina and Canso also rank among the amphibian scooper planes.
These Type III airtankers, made both in Canada (Canso) and the USA (Catalina), are used for firefighting not only in the USA. The picture shows Airtanker 85 (# N85U) of the US Flying Fireman charter company.

A specialty of the PBY is the observation bulge toward the rear.

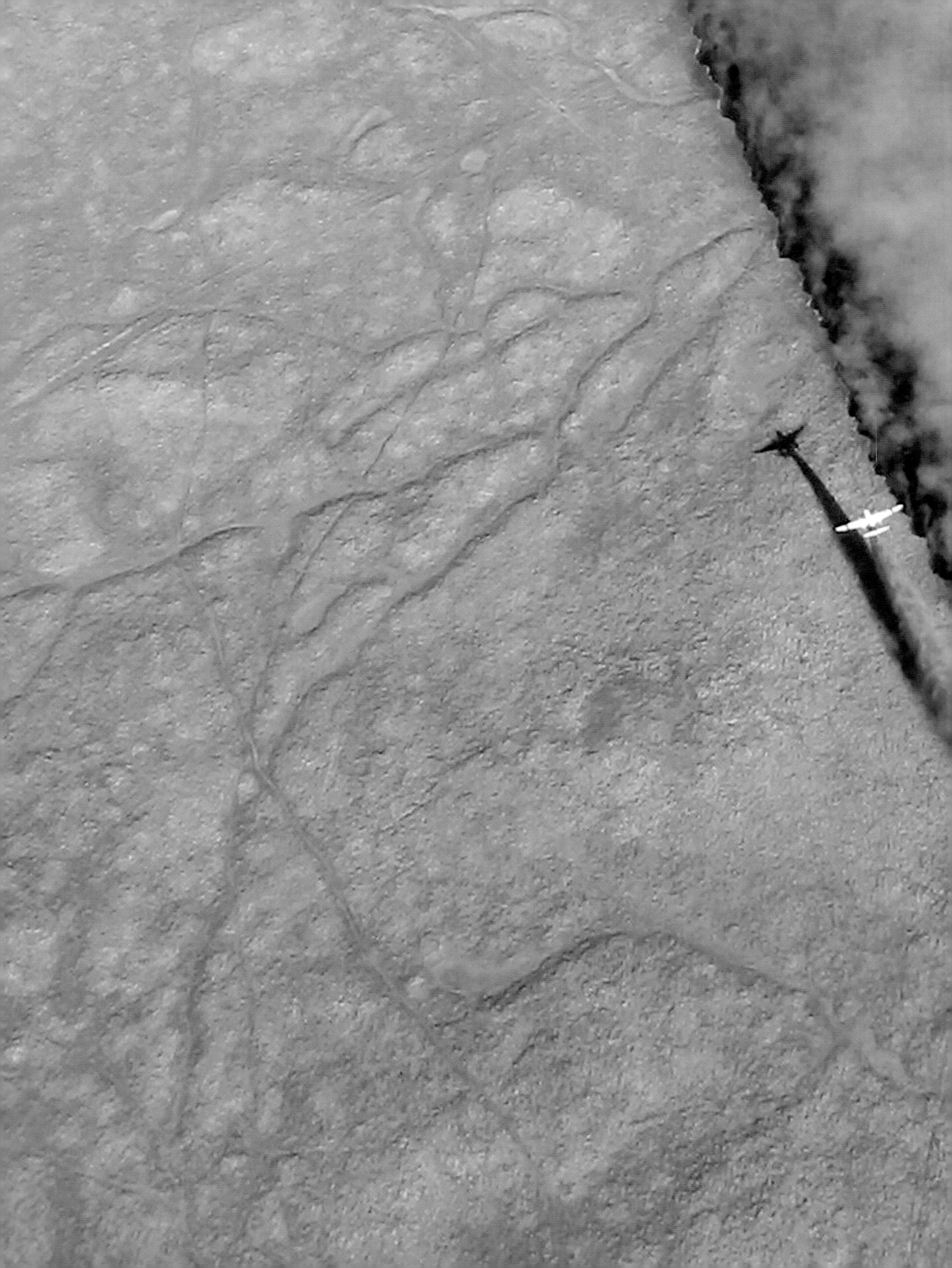

The Future of Aerial Firefighting

The Future of Aerial Firefighting

Air support will still be indispensable in the field of forest and surface firefighting in the future.

See also:
Aerial Firefighting—The History of Firefighting Aviation.

A short look backward in history: The pioneers of aerial firefighting in the USA date from the years after 1950. They were, among others, Granville Swift, Robert Semple, and Floyd Nolta, who first brought new theories and technologies of firefighting up for discussion in the realm of agricultural aviation. Joe Ely of the U.S. Forest Service (USFS) instituted considerations in 1955 as to using aircraft for fighting forest and surface fires. On August 13, 1955, the first dropping of water on the Mendenhall Fire in the Mendocino National Forest in California took place. Vance Nolta, Floyd Nolta's brother, flew a Boeing Stearman Caydet, a modified agricultural plane. This plane, numbered N75081, was the first registered airplane in the history of forest fire aviation.

Later, private American fire-protection groups recruited mustered-out World War II warplanes and transports from the U.S. Air Force and equipped them with water tanks and dropping equipment. The best-known manufacturers were, and still are today, the Aero Union Corporation of Chico, California, Hawkins and Powers of Greybull, Wyoming, Minden Air Corporation of Minden, Nevada, and Neptune Aviation of Alamogordo, New Mexico.

The most varied military airplanes were used, and were divided into size classes I to IV: Type I airtankers with more than 3000 gallons of retardant, Type II airtankers with 1800 to 2299 gallons, Type III airliners with 600 to 1799 gallons, and Type IV airtankers with 100 to 599 gallons of retardant (100 gallons = ca. 235 liters).

Among the large airplanes of Types I and II, which were used until the U.S. Forest Service (USFS) decided to retire them in May 2004, were the Lockheed C-130A, Lockheed P3-A Orion, Douglas DC-7 (all Type I), and Type II airtankers including the Douglas DC-6, Lockheed P2V, SP2H, PB4Y2, and Douglas DC-4.

These large airtankers used in the USA were generally in service for an average of 50 years (48 to 60). Despite their regular servicing and checking by charter contractors, both their technical vulnerabilities and repair problems, due to a lack of spare parts, increased. Despite all safety precautions, there were several serious accidents and crashes with total losses of the planes plus fatalities.

The safety of veteran airtankers was discussed intensively, as was general safety in aerial firefighting, in 2002 and 2003. In April

2004, the U.S. Forest Service (USFS), the largest American national forest and forest fire agency, in agreement with the National Transportation and Safety Board (NTSB) and other national forestry and forest fire (DOI) agencies, decided to immediately ground the 33 large airtankers (Types I and II) then in use. With that, aerial firefighting faced the greatest crisis in its history, as well as probably the greatest technical and tactical challenge for the future in terms of a medium-length National Fire Aviation Program.

Meanwhile, some of the large airtankers in the USA were temporarily put back into service. After very thorough inspections and an extensive technical analysis by experts of the Dyna-Corp Technical Service, the U.S. Forest Service (USFS) and the Bureau of Land Management (BLM) accepted the limited-time use of seven Lockheed P-3A Orion (Aero Union Corporation) and nine Lockheed P2V Neptune (Minden Air Corporation, Neptune Aviation) airtankers during the 2007 forest fire season. All the planes were fitted with Traffic Collision Avoidance Systems and Operational Load Monitoring Equipment (OLME), a system for monitoring retardant loading and dropping.

The appropriate contracts were set up as "call-when-needed contracts" (CWN), thus giving priority to the other aircraft used in fighting forest and surface fires (Type III airtankers, SEAT, Type I-III helicopters).

Through financial support programs, numerous additional fire helicopters and helitankers of Type I (6 in 2005) and Type II (24 in 2005), and about 100 Single Engine Air Tankers (SEAT) were chartered or bought for forest and surface firefighting.

The California Department of Forestry and Fire Protection (CDF) and the U.S. Forest Service (USFS) tested a possible future project in California under the name of FireHog™. In it, A-10 Thunderbolt II ex-military fighter planes were considered for use as airtankers and as successors to the S-2T (CDF Type III airtankers).

Several large aircraft manufacturers and charter companies were also active in matters of future aerial firefighting. Soon after the grounding of the large airtankers, the first concepts for the development of especially large and high-performance firefighting planes (supertankers) and helicopters were discussed. The Evergreen International Aviation Firm, Inc. of McMinnville, Oregon, thus undertook in 2006 the first flight and drop tests with a Boeing B0747, which was equipped with a 24,000-gallon (ca. 90,850-liter) tank, and the Minden Air Corporation, a charter firm long involved in forest and surface firefighting, equipped the four-jet British Aerospace BAe 146-100 short-range passenger plane as an airtanker. Other passenger planes were at least planned for technical modification (conversion) to firefighting planes, such as the Lockheed L-188 Electra, Boeing B-727, Boeing DC-10, and Boeing B-737. A Bombardier Dash-8 Q300 has already been rebuilt and put to use as an airtanker by the Neptune Aviation firm of Missoula, Montana—and another Dash-8 (Milan 73) is used for firefighting by the

See also:
Fatal End of Service—The Crash of Fire Plane 130—Airtanker crashes in forest fire service, and The End of the Large Airtanker in the USA.

French Securite Civile. From the military realm, 25 Bell AH-1 Cobra (Model 209) helicopters were rebuilt into lead and command planes by the U.S. Forest Service (USFS) and have already been used successfully. The same is true of the UH-60 Black Hawk, which is used by, among others, the Los Angeles County Fire Department in California as a Type I Helitanker.

These few examples show that the future of aerial firefighting has begun, at least in the USA.

Finally, let us look forward into the future of Europe and the rest of the world, and add a personal note. It should have become clear already that forest and surface firefighting from the air will change decisively, in the short and long terms, in regards to the previously used technology and tactics. The heyday of the large airtanker may be mostly at an end.

This is particularly true of the USA, where the large airtanker has for decades played a major role in fighting forest and surface fires. Now we will see what aircraft technology will prevail after the final retirement of the Type I and II airtankers. As a result of the initial estimates, it will probably not be the super-airtanker—perhaps a more likely concept is a high-performance middle-sized airtanker. Smaller but more potent airplanes (Type III airtankers, SEAT) and capable helitankers (Type I and II helicopters) will finally win the "race" over the American woods and wildlands, for both economic and tactical considerations.

In Europe—and especially the Mediterranean area—the large airtanker never really established itself in the realm of forest and surface firefighting from the air. From the beginning, it was the medium-sized (Type III) amphibian planes that dominated tactical firefighting. The Single Engine Air Tanker (SEAT) and large and medium fire helicopters offered purposeful tactical support to the ground units. Here too, the future may be clear: Canadair CL-415 amphibian firefighting planes, Single Engine Air Tankers (SEAT) and Type I and II helitankers, in the end, offer optimal aerial firefighting—a concept, moreover, that has been practiced in a similar form for years in Canada.

In the Russian states, forest and surface firefighting from the air will presumably not change much from the present situation in the near future. Airplanes of the US Types I to III, high-performance helicopters (including helitankers), and various other aircraft (such as crew helicopters for rappelling and jump planes for smokejumpers), are very reminiscent of the American system of aerial firefighting. Just recently, the Russian Ministry for Civil Defense and Catastrophe Protection obtained seven more Ilyushin IL-76 TP large airplanes (in US terms, Type I airtankers) for the state forest and forest fire organization, Avialesookhrana.

Otherwise, most countries in all parts of the world are banking extensively on the use of fire helicopters (of all sizes, with more Type I and II copters such as the Sikorsky S-64A Skycrane) as well as Single Engine

Air Tanker (SEAT). To what extent the newest technical considerations, concepts, and developments will prevail can scarcely be said at this time. Zeppelins (Wetzone Engineering of California) or Firebirds (ROA. UAV) from Israel Aircraft Industries definitely will not do the job of fighting forest fires—nor is it likely the planning principle that has come to the fore after several years of major forest fires – "The more extinguishing water, the better!" – for tactical reasons, turn out to be the last word in firefighting wisdom.

See also:
The Future of Aerial Firefighting—Conversion, Development, and Rebuilding of Aircraft for Forest and Surface Firefighting.

An Air Tractor AT-802F Fire Boss amphibian plane, one of the solutions for the future.

A Fatal End of Service—The Crash of Airtanker 130

About 500 inhabitants live in the small town of Walker, some 15 miles (24 km) west of the border between the states of California and Nevada. The town lies at the foot of the northern Sierra Nevada, 75 miles (120 km) south of Reno.

The peace and calm in Walker changed very quickly on that Saturday, July 15, 2002. At the headquarters of the Sierra Front Interagency Fire Center in Minden, Nevada, the report of a wildfire came in about 12:00 noon, a forest and surface fire in the undeveloped and natural vegetation. The weather conditions on that day were extremely unfavorable and the danger of a forest and surface fire was high. For days people had complained about the unending dryness, the extreme heat, the very meager moisture in the air, and the warm, gusty winds.

The responsible fire managers of the U.S. Forest Service (Humboldt-Toiyabe National Forest, Bridgeport Ranger District) of the Bureau of Land Management (BLM), the Bureau of Indian Affairs (BIA), the Nevada Division of Forestry (NDF), the California Department of Forestry and fire Protection (CDF), and the local fire departments immediately sent fire trucks and fire crews to the scene, not far from the village of Walker. Within a short time, some 1550 firefighters with numerous vehicles, bulldozers, six fire helicopters, and two airtankers, were in action at the forest and surface fire now called the "Cannon Fire."

The attack was led by an Incident Management Team (Type 3) under the direction of Bob DeBaun. He and his experienced team succeeded in bringing 85% of the fire under control within the next few days. In all, an area of some 22,700 acres (about 92 square kilometers) on both sides of Route 395 had been affected. As a precaution, the eastern parts of the town had to be evacuated for a time, and some 200 inhabitants were requested by the Mono County Sheriff's Department to leave their homes.

The Cannon Fire was probably caused by members of the U.S. Marines, who had held survival training in an isolated part of the Humboldt-Toiyabe National Forest and set campfires there.

On Friday, June 29, 2002, the Cannon Fire was finally brought under control. While 22,750 acres of wood and brush vegetation had burned, the firefighters were able, by laying fire lines and setting back fires, to protect almost the whole town of Walker. Despite the sometimes considerable danger of the 300-meter wall of flames reaching into the village of Walker, the fire destroyed only one vacation house, one vehicle, one garage, and two small wooden sheds. The costs of the Cannon Fire added up in the end to some 7.9 million dollars.

Airtanker to Action!

The scene changes, it is Monday, June 17, 2002, about 2:00 P.M. Steven Wass, 42, a longtime airtanker pilot, his co-pilot Craig LaBare (36), and flight engineer Michael Davis (59) prepared their plane, a Lockheed C-130A, for service at the Cannon Fire. The Hercules with number N130HP, a former Air Force transport, was stationed during the forest fire season at the Minden-Gardnerville Airtanker

The crew of Airtanker 130: Flight Engineer Tony Griffin, Pilot H. F. "Buzz" Schaffer, and the late First Officer Craig LaBare, who lost his life along with his crew in the crash of the airtanker.

Base south of Reno, Nevada—only some 50 miles away from Walker, California.

The airtanker bearing number 130 was one of the largest firefighting planes allowed for forest and surface firefighting in the USA, and was chartered by the Forest Service of the U.S. Department of Agriculture from the Hawkins and Powers Aviation firm (H&P) in Greybull, Wyoming, for the duration of the forest fire season.

The C-130A Hercules was a Type I airtanker according to the ICS (Incident Command System) classification, equipped with a 3000-gallon (11,355-liter) retardant tank. Via eight hydraulic hatches, the retardant, a mixture of water and certain chemicals for cooling the burning areas, could be dropped from the air. The wingspan of the big plane measured 133 feet (40.5 meters), its length was almost 100 feet (30.4 meters). Its top speed was about 275 mph (ca. 442 kph). The C-130 Hercules planes were also used by the U.S. Air Force and National Guard for forest and surface firefighting. They were equipped with a Modular Airborne Fire Fighting System (MAFFS).

See also:
Airtankers in the USA—The Modular Airborne Fire Fighting System (MAFFS).

130
HIGH
INTENSITY NOISE

Waiting for the next call! Airtanker 130 waits at the Libby Airtanker Base in Fort Huachuca, Arizona.

It is shortly after 2:30 P.M. that Airtanker 130 is over Route 395 in California en route to the scene of the fire. Pilot Steven Wass steers the plane, loaded with 3000 gallons of retardant, toward the left flank of the fire north of the town of Walker and goes into a descent. A few meters over the fire, the T-130 crew opens the hatches of the tank and lets the retardant drop in a full flood over the fire area.

See also:
Airtankers in the USA—Type I Airtankers.

At that moment the unthinkable happens. The two wings of the plane suddenly fold upward and break off—the plane drops sharply to the ground and crashes into the wooded hills not far from Route 395. It is 2:45 P.M. when Steve Wass, Craig LaBare, and Michael Davis die practicing their profession as wildland firefighters.

The crash is a shock for the crews working on the Cannon Fire. Many members of the crew working close to the fire line had watched the airtanker's flight and directly experienced the simply unbelievable event. Hours later they still stand numbed at the scene and mourn for their comrades.

A Personal Loss

As I write this report, I remember the days at the Libby Airtanker Base in Fort Huachuca, Arizona, in July 2001. The small airfield of the U.S. Forest Service (USFS) lies right by the Mexican border and in the midst of a U.S. Air Force base; Airtanker 130 was stationed there during the 2001 forest fire season. During my work on the subject of "forest and surface firefighting," I wanted to visit the southernmost fire airtanker base in the American West. The late Air Service Manager, Bill Parks, Chief of the Huachuca aerial firefighters, expected me and had already prepared my quarters for a stay of some days.

I was surprised when I saw the T-130 on the runway. I had absolutely not expected to see one of the largest forest firefighting planes in American air attack service at the dry, dusty desert region on the Mexican border.

And so I met the crew of T-130: Pilot H. F. "Buzz" Schaffer, First Officer Craig LaBare (†), and Flight Engineer Tony Griffin. All three were longtime, experienced firefighting aviators, with tangible excitement about their job and their aircraft. And they shared their excitement with me gladly and generously. It was "Buzz" Schaffer who told me in detail about the air attack system and the crew's work in the ongoing forest fire season, and Craig LaBare showed me every imaginable detail of "his" airplane. He made time for me, and I enjoyed getting to know the plane and the private living space of the aviators within it: the cockpit, the navigator's workplace, the joint "living and sleeping room" of the crew inside the gigantic fuselage. And not least, there was the feeling of being allowed to sit in the pilot's seat in the cockpit. It was really great.

One could also sense that the crew had faith in their plane. "Reliable, easy to control despite the weight, fast!" said Pilot "Buzz" Schaffer.

I had a good and interesting time with the T-130 crew at Libby Airtanker Base, and I profited for a long time afterward from the wealth of impressions and factual information.

The days in Fort Huachuca went by much too fast. When I parted from Craig LaBare, "Buzz" Schaffer, Tony Griffin, and Bill Parks, it never would have occurred to me that barely a year later this splendid machine could crash so terribly on a firefighting call. Craig LaBare lost his life, along with his partners Steven Wass and Michael Davis. The news that I heard from wildland firefighter friends in the USA had shocked and saddened me. Craig LaBare, Steven Wass, and Michael Davis had lost their lives in practicing their profession of fighting forest and surface fires. They died, like many of their colleagues before them, because their job is one of the hardest and most dangerous jobs in this world.

Dennis Faber of Greybull, Wyoming, formerly chief flight engineer at Hawkins & Powers, had flown Airtanker 130 during the 1996, 1997, and 2001 forest fire seasons and just two weeks before its crash. "She was a good ship, and Steve, Craig, and Mike were all good friends and good men."

After the crash of Airtanker 130, a camera team of the German Fokus TV television company came from Munich to see me. It was about a planned broadcast on the event at Walker, California, about the crash, and about forest and surface firefighting in the USA in general. At that time, the background of the crash could be articulated only very cautiously. "It may involve the fact that the retardant, several tons of liquid, was suddenly dropped, and at that moment the weight relationship came into effect. The plane may thus have been overstressed."

Author Wolfgang Jendsch during a television report (Fokus TV) on the crash of T-130.

Meanwhile the National Transport Safety Board (NTSB) of the USA had concluded its investigation of the cause of the accident. Several cracks in parts of the wing structure, obviously resulting from metal fatigue, were found.

After that, all Lockheed C-130A firefighting planes were grounded at once. Except for the C-130s equipped with the military Modular Airborne Fire Fighting System (MAFFS), they would never again see service over the burning forests of the USA.

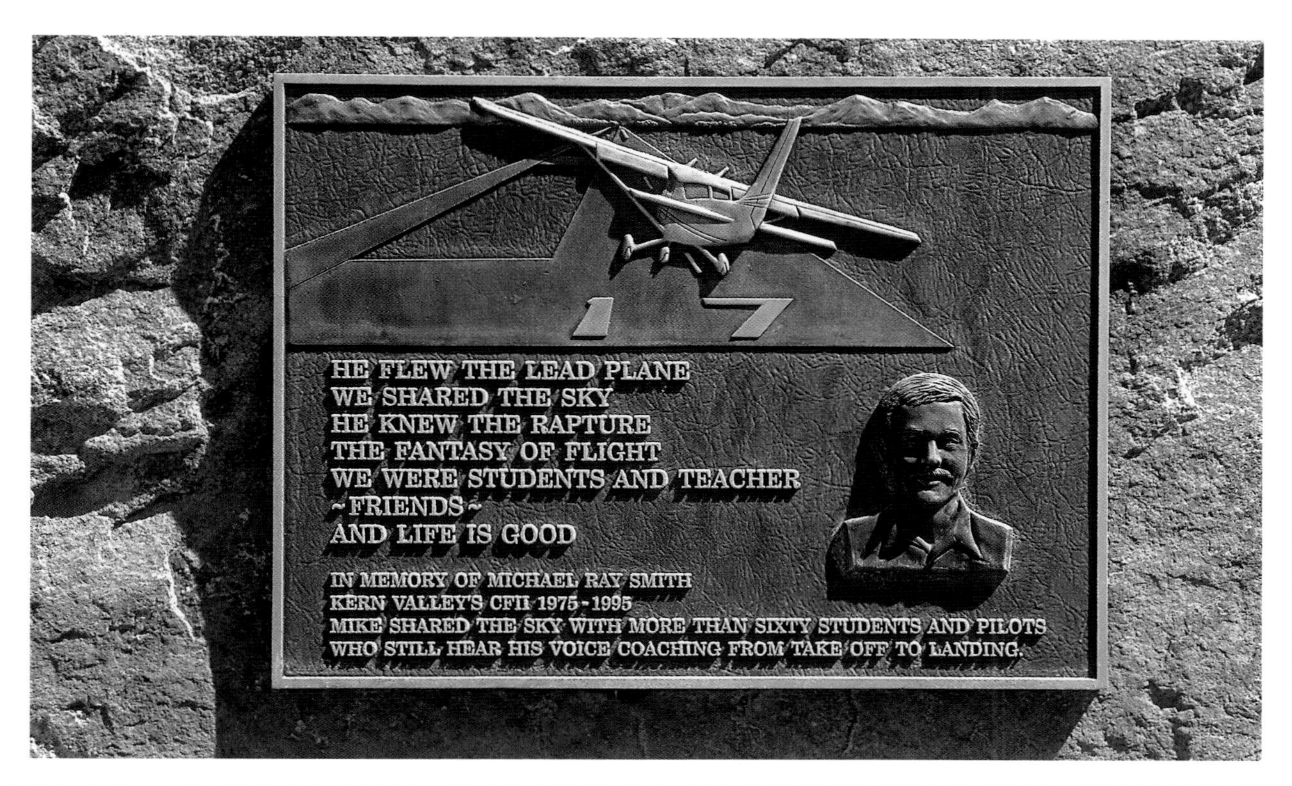

Again and again, families and friends of crashed and killed flying firefighters must remember—as on this memorial plaque in Kern County, California.

Visible signs of danger with which firefighters must live are the wrecks of their crashed aircraft. The picture shows a Bell UH-1D of the Nevada Division of Forestry (NDF) that crashed into a mountain lake in the Sierra Nevada Mountains. Good and bad luck: Six members of a fire crew remained uninjured, one was just slightly injured. The reasons for the crash were, among others, stated after investigations of the National Transport Safety Board (NTSB) of the USA: The individual action conditions (height, air, temperature, load) in the high mountains.

The work of the wildland firefighters on the ground, and particularly those in the air, is especially hard and dangerous. Perhaps for that very reason, inhabitants of the fire-threatened woods and wildlands are especially thankful to the firefighters for protection and help in emergencies.

Examples of Airtanker Accidents in Fighting Forest Fires

The fighting of forest and surface fires is dangerous. Many firefighters have had to pay with serious injuries, or even with their lives, every year for their involvement in the battle against the forces of nature. Fighting these fires from the air (aerial firefighting) is even more dangerous. The number of crashes or near misses, especially among firefighting aircraft, has increased considerably in the last few years. The dangers of fire, smoke or turbulence right over the scene are not alone to blame; so is, above all, the techniques used in action.

See also:
The End of the Large Airtanker—the debated decisions of May 2004.

The American U.S. Forest Service (USFS) has, on request from the National Transportation Safety Board (NTSB), grounded all "large firefighting planes" (Type I and II Airtankers) in the USA in May 2004, for safety reasons.

Here is a series of serious flight accidents in aerial firefighting that have been investigated in terms of accident technology and evaluated by the National Transportation Safety Board (NTSB) and the involved fire agencies.

CDF Lead Plane Crashed

In September 2006, a Rockwell OV-10 Bronco lead plane (Air Attack 410) of the California Department of Forestry and Fire Protection (CDF) crashed in Mountain Home Demonstration Forest not far from Springville in Tulare County, California. The two occupants, pilot George "Sandy" Willett (52) of Hanford and CDF battalion chief Robert Paul Stone of Visalia, lost their lives in the crash.

The plane with registration N419D was on an observation flight some 2 miles north of Porterville. For unknown reasons, it touched the tips of some of the 125-foot (ca. 38 meters) trees at the edge of a steeply sloping rim of a canyon—a latent danger even in real forest fires. The crash of the plane was caused by this contact.

C-130 Airtanker Crashed in France

In September 2000 a Lockheed C-130A Hercules of the American contractor T & G Aviation crashed after a flight maneuver northwest of Aubenas, France. The Type I airtanker (Airtanker 82, built in 1957), which was stationed at the Marignane fire airbase near Marseilles during its contracted stay in France, was in forest firefighting action under contract with the French Securite Civile. The crash, in which

Lead plane 410 (AA 410, # N419DF) crashed on September 6, 2006.

two of the four crewmen were killed, took place after water was dropped. The plane then went out of control and crashed into a hill.

Helicopter Crash Took Four Lives

In August 2001, the crash of a Eurocopter AS-350 B3 took the lives of three wildland firefighters and the pilot. The copter was under contract to the U.S. Forest Service (USFS) and on the way to the Krassel guard station in Payette National Forest, Idaho, when it went into a spin and crashed about 18 miles west of Yellow Pine.

Crash of Airtanker 130 Near Walker, California

Metal fatigue on the wing attachments of the 45-year-old Lockheed C-130 Hercules airtanker (# N130HP, Hawkins & Powers) was the reason for the crash of the Type I Airtanker 130 at Walker, California, in June 2002. The wings of the plane broke off after the plane had dropped its fire retardant. Pilot Steven Wass (42), co-pilot Craig LaBare (36), and flight engineer Michael Davis (59) lost their lives.

See also:
The End of the Large Airtanker—A Fatal End! The Crash of a C-130.

Airtanker 130 (# N130HP) crashed on June 15, 2002.

Russian Fire Helicopter Crashed in Turkey

In August 2006, a Russian Kamov KA-32 fire helicopter crashed in the Lake Gombe area of southern Turkey. The copter was about to pick up water for firefighting when it went out of control and plunged to the ground. The five-man crew of the helicopter was lucky; the Turkish pilot, his co-pilot, and three Russian helpers survived the accident with moderate injuries.

US Airtanker 99 Crashed

In October 2003 a Lockheed P2V Type II airtanker (Airtanker 99, # N299MA, Minden Air Corporation), struck rocky land near the East Highlands of California (Bear Lake) in a flight maneuver and crashed. Both crew members lost their lives in the crash. The crew of a fire lookout had already observed smoke and noticed that the plane then made a "rather abnormally steep" 180-degree curve before it went out of sight.

Wing Broke Off After Explosion

In August 1994 a Lockheed C-130A (Airtanker 82, # N135FF, Hemet Valley Flying Service), chartered by the California Department of Forestry and Fire Protection (CDF, now CAL FIRE) and the U. S. Forest Service (USFS), left the airtanker base at Hemet, California, on a forest fire call. About 20 minutes later, eyewitnesses saw a fiery orange glow on the plane's right wing over the Tehachapi Mountains. Only a few seconds later another red fireball and thick black smoke were observed. The right wing broke off the plane. Right after that, it vanished behind a range of hills and crashed not far from Pearblossom, California. The airtanker was completely destroyed, and three crew members were killed in the crash.

Investigations by the National Transportation Safety Board (NTSB) revealed later that the airtanker's fuel system had obviously exploded and caused the wing to be torn off.

Crash in Mountainous Terrain

Airtanker 26, a Lockheed P3-A Orion (# N76AU, Aero Union Corporation), was on a training flight from the Chico Air Attack Base to the Ishi Wilderness some 30 miles northeast of Chico in northeastern California in April 2005 when it crashed in trackless terrain of the Lassen National Forest. The three crew members, Aero Union chief pilot Tom Lynch, Paul Cockrell, and Brian Bruns, were killed in the crash.

Whether a pilot's error was the cause of the airtanker's crash was unknown for a long time. No longer controllable flight and firefighting maneuvers in narrow and mountainous terrain resulted in the accident.

A Personal Word

These and many other accidents with firefighting planes are saddening, for firefighters lost their lives in the course of their hard and dangerous work of fighting forest and surface fires—also saddening because I had gotten to know some of these subsequently crashed airtankers (such as Lead Plane 410, Airtanker 130, Airtanker 99, and Airtanker 26) and their crews personally during my numerous visits, had sat with them in the cockpit, and had the opportunity to get to know the technology and tactics of these fascinating airplanes thoroughly. I have also been able to make many a flight in such planes. Thus I think back with sadness to those experiences and will preserve their memories in honor of those who lost their lives doing their duty.

Airtanker 99 (# N299MA) crashed on October 3, 2003.

26

Airtanker 26 (# N926AU) crashed on April 20, 2004.

The End of the Large Airtanker in the USA

See also:
The Future of Aerial Firefighting

The fighting of forest and surface fires, and aerial firefighting in particular, are dangerous. Fire managers and firefighters have long known this and try to counteract the potential dangers through personal qualification, physical ability, forward-thinking tactics, and extensive safety measures. Despite this, there have been tragic and sometimes fatal accidents time after time.

On June 17, 2002, the Airtaker 130 (# N130HP), a 46-year-old Lockheed C-130A of the Hawkins & Powers firm, crashed while fighting the "Cannon Fire" near Walker, California. The three-man crew lost their lives. After later investigation by the National Transportation Safety Board, it was found that several cracks in the wing structure, obviously resulting from metal fatigue, were the cause of the accident. In 1994 a C-130 had crashed under similar conditions.

At the end of June 2002, the left wing of a 50-year-old Consolidated-Vultee PB4Y2 airtanker (# N7620C) of the Hawkins & Powers firm broke in flight during a firefighting action near Estes Park, Colorado. The plane crashed immediately; the crew—pilot and copilot—died. Here too, metal fatigue was later determined to be the cause.

During a return flight from a forest fire in Arizona, Airtanker 99 (# N299MA), a twin-engine P2-V of the Minden Air Corporation, crashed in San Bernardino National Forest, not far from San Bernardino, California, on October 3, 2003. Both crew members were killed. Despite these striking events in aerial firefighting in the USA, the news on May 10, 2004, struck like a bomb: The U.S. Department of the Interior (USDI) and the U.S. Department of Agriculture (USDA) announced that the contracts for 33 large airtankers of Types I and II "in view of the incalculable technical condition of the aircraft and in view of public safety," were cancelled at once. The U.S. Forest Service (USFS), America's largest national forest and fire protection agency, referred to an investigation report of the National Transportation Safety Board (NTSB) of April 23, 2004, in which a conclusion drawn from the investigations of previous crashes of airtankers in 1994, 2002, and 2003 were cited.

"It is obvious that at this time no effective technology is available that will assure the future ability of these firefighting planes to fly," thus spoke the NTSB. At the same time it was learned that both the U.S. Forest Service (USFS) and the Department of the Interior (USDI) are responsible for the maintenance of the safety of firefighting aircraft—and thus also for the safety of the flying personnel and the ground units.

Most of the airplanes used as large airtankers were developed for military use decades ago, before they had been mustered out, taken over by private firms, and rebuilt into firefighting planes. For the annual forest fire season in the USA, the planes were then chartered by the fire agencies, including the U.S. Forest Service (USFS), the Bureau of Land Management (BLM), the Bureau of Indian Affairs (BIA) or the California Department of Forestry and Fire Protection (CDF) from the contracting firms that specialized in aerial firefighting.

"For the fewest airplanes, though, a faultless technical service and regular inspection could be documented," the NTSB declared in their investigation. This "faulty evidence" made it impossible to guarantee sufficient technical flight safety for the airtankers in the future. After all, the large airtankers were from 48 to 60 years old!

"Safety is the highest goal in fire protection, and thus indispensable for us," the director of the U.S. Forest Service (USFS), Chief Dale Bosworth, declared from the USFS National Headquarters immediately after the NTSB investigation became known. (Tip: The leadership of the U.S. Forest Service had changed in 2007. The new Chief of the national forest and forest fire agency is now the former regional forester from Missoula, Montana, Gail Kimbrough. Also Kathleen Clarke, Director of the Bureau of Land Management (BLM), agreed with this decision: "A continuation of the use contracts for these large airtankers constitutes a non-acceptable risk for those responsible for air traffic, for the crews on the ground, and for the communities that we have to protect. Therefore we take the investigations and the concluding recommendations of the NTSB very seriously," said Kathleen Clarke. Both organizations made clear that efforts would be made in cooperation and agreement with the other state fire protection organizations, to make clear to Congress, the industry, and the public that this decision became necessary to be able to assure safety, effective fire management, and effective fire protection.

The USFS and BLM also stated that the use of large airtankers "is only one of numerous possibilities that action forces can use in fighting forest and surface fires. Such actions are generally directed and carried out from the ground—not from the air! In the course of one year, thousands of forest and surface fires are extinguished in the initial action—without air support!" the U.S. Forest Service (USFS) argued.

For the 2004 forest fire season, the fire agencies first developed an immediately applicable strategy by which the ground firefighting units (hotshot crews, engine crews, etc.) should be supported by the coordinated action of other available aircraft—such as fire helicopters and helitankers (Types I to III fire helicopters), Single Engine Air Tankers (SEAT), and the eight C-130 National Guard and Air Force planes equipped with the Modular Airborne Fire Fighting System (MAFFS).

The planning for "long-term aviation management" and the procurement of modern aircraft—dependent on the available financial means—will now be done under the leadership of the U.S. Forest Service (USFS/USDA) and the Department of the Interior (DOI) organizations (BLM, NPS, BIA, OAS).

"Thereby it is clear that the service time of older aircraft with unclear technical flight capability for forest fire action is past," said Chief Dale Bosworth. "We are thankful to the pilots, the crews, and the operators of these aircraft that they have been involved in the use of airtankers for forest and surface firefighting. After all, we in the USA have the greatest fire management and the best fire protection concepts in the world, and we will also continue to be concerned about protecting human life, property, and the natural resources of our nation."

Previously, the leading fire protection organizations had spent some 30 million dollars a year on the use of airtankers.

The Lockheed C-130 (MAFFS) planes of the U.S. Air Force and U.S. National Guard were to stay in use, as were the Type III S2 and S2-T airtankers of the California Department of Forestry and Fire Protection (CDF) and the Single Engine Air Tankers (SEAT).

The remaining 33 large airtankers, according to the USFS and BLM, were to be replaced for seasonal work by up to 80 other airtankers (SEAT) and 71 Type I and II helicopters. Some 66 million dollars had to be found at the start to pay for these additional units.

Also available—besides the 22 CDF airtankers already noted—were the numerous helicopters (using buckets) and helitankers (with fixed tanks) that were chartered annually by the fire agencies for forest fire use. The Los Angeles County Fire Department (LACoFD) had then signed contracts to lease Canadair CL-415 (Super-Scooper) planes to protect their area of responsibility, with its 51 cities, more than nine million inhabitants, and a total area of 2200 square miles (5700 sq.km.). The LACoFD had first used Canadair firefighting planes during the 1994 and 1995 forest fire seasons. The fire managers in Los Angeles would not buy the Super-Scoopers, since the planes were not optimally suited for the regional conditions, and were also very expensive.

The attempt of San Diego County to make U.S. Navy and Marine helicopter pilots "fit" for forest firefighting in quick courses was not very promising either. The war in Iraq prevented the continuing use of military aviators. "As usual," CDF Battalion Chief Ray Chaney summed it up after a training course for pilots of the 3rd Marine Aircraft Wing from Miramar and the HC-85 Navy Reserve Squadron from North Island, "these pilots cannot yet be certified for forest fire use until they have trained sufficiently long and intensively with CDF crews. This in turn is not possible until the CDF forest fire aviators have finished time-consuming military training."

Discussions at All Levels

Understandably, the waves rose high when the decision of the U.S. Forest Service to do without the 33 large airtankers became known. Misunderstanding in professional circles, plus protests from groups of pilots and wildland firefighters immediately characterized the discussions at the airbases and fire stations of the individual fire-protection organizations, and particularly the reporting in the media of almost all US states. It was not so much the displeasure about lost jobs that caused the protests as it was primarily the evaluation of experienced work forces, which could not be done without on large airtankers in the fighting of forest and surface fires.

"The large fixed wings were not grounded, they were terminated!" was the emotional outcry of an airtanker pilot, who also raised the question of responsibility: "Who is going to be responsible now for the individual foreman on the ground who loses his life because air support in a dangerous fight against the flames was not available?" "I simply cannot understand the decision of the Forest Service and the Department of the Interior," another fire manager sighed, asking: "If this decision was really necessary so soon after the beginning of the active forest fire season, why don't they promise us a sensible transition phase?" "We will have a season without our guardian angels," said another voice. "How many lives will be lost, how many homes will be lost, how many families and businesses will be ruined?"

"It hit here like a bomb," said airtanker pilot Phil Darnell (63) of the Prescott Fire Center in Arizona. When he heard on the Monday evening after the USFS decision that he had lost his job, he felt "sick to my stomach." "I had reckoned on conceptual changes, but that they'd cross out everything, I had never expected that." At the time, Phil Darley flew the ex-Navy Lockheed P3-A Orion from Prescott back to Aero Union's home base in Visalia, California, one last time, while his colleague Del Hunt flew his 1947 Douglas DC-6 to the TMT Butler Base in Chico. Two SEATs would then replace the fire-bombers in Prescott.

The criticism of the U.S. Forest Service's decision also showed political ramifications. A non-American airtanker pilot wrote to his colleagues in the USA: "I have worked at this job more than twenty years, and I cannot understand why the USA is ready to spend more money in one week in Iraq than to finance an airtanker fleet to protect your houses and property."

Along with the emotional reactions, there were also professional voices that spoke against the decision of the U.S. Forest Service (USFS). Fire managers doubted that the fire-protection organizations would be able to survive the year's tough forest fire season successfully. Criticism was also directed against the announcement from the U.S. Fire Service (USFS) to involve more military MAFFS in aerial firefighting. "Such a plane costs $7000 to $8000 an hour," said a P2V pilot, and "with all respect for the crews, MAFFS cannot replace the formerly used airtankers." For their pilots, fighting forest fires was not the highest-priority task among many others. Thus the use of the MAFFS could only be an alibi for the Forest Service. In terms of the annual cost of aerial firefighting, specialists pointed to the resulting costs if smaller forest and surface fires were not extinguished directly through air support and could spread dangerously.

Responsible USFS personnel opposed the presumably unexpected and critical reactions verbally. "The war against the fire is over!" one of them declared; "we will no longer fight fires, we will manage them!" The answer from a California firefighter arrived by return mail: "And while you manage the fires, I hope they will not take our S-2 airtankers from the initial action base to extinguish the managed fire."

One fear was expressed clearly in California: "They will leave our citizens and taxpayers in the Wildland Urban Interfaces (widespread settlements in wooded and natural regions) without air support in forest and surface fires"—in fact a life-threatening situation if forest and surface fires spread in the vicinity of settlements. Fire managers also doubted that the substantially smaller SEATs were capable of replacing the large airtankers. Neither the loading capacity of extinguishing materials nor the qualifications of the pilots, who were previously used essentially to spray agricultural fields, would suffice. A manufacturer of these small planes, Air Tractor Inc. of Olney, Texas, naturally saw this differently. The firm vigorously promoted the capability of their newest product, the AT-802 F, with a load capacity of 820 gallons (ca. 1930 liters). The helicopter situation looked similar: Type II and III copters have only limited capacities; Type I helicopters are only partial alternatives but are available only in small numbers.

See also:
Airtankers in the USA—Single Engine Air Tankers (SEAT)

"Airtankers are an indispensable part of our national efforts in the realm of forest and surface firefighting," said Kristen Schloemer, the president of Neptune Aviation in Missoula, Montana, one of the largest and best-known airtanker contractors in the USA. "Other airplanes cannot fill these gaps," Schloemer declared. Her business made up to eight large airtankers and their highly qualified crews available for U.S. Forest Service (USFS) use every year. Schloemer also contradicted the opinion that the large airtankers were too unsafe for forest fire use. "We have servicing programs that go beyond those of other aviation firms and set qualitatively higher standards than those of many airlines," was her evaluation. Despite massive emotional and professional criticism from almost all sides, there were other opinions. "I support the decision of the government organizations completely," Ray Weidenhaft, Fire Management Officer of the Wyoming State Forestry Division, declared. Weidenhaft knows what goes on in aerial firefighting: "The responsible parties cannot bear these risks." "But," the fire manager went on, "I cannot believe that it will not hit us much harder in the future, because there will not be enough helicopters and Single Engine Air Tankers (SEAT) available to replace the withdrawn airtankers."

"The sky is not falling," said Dan Torrence, Manager of the Redmond Air Center in central Oregon. "Airtankers are not our only weapon against the fires," said the forest fire expert, who saw possibilities in compensating for the loss of the large airtankers with other aircraft. In this context, Torrence referred specifically to the average age of the ex-World War II planes of 46 years and thus the probability that the technical safety of these airtankers could not be guaranteed on account of their age. "We will probably have to wait for the next generation of modern airtankers," said Rex Holloway, spokesman of the U.S. Forest Service (USFS) in the organization's regional office.

Other fire protection organizations and units in the western USA prepared "silently" for a forest fire season. "We want to maintain safety here," Gail Aschenbrenner of Coronado National Forest in Arizona promised. Preventive fire protection was stressed there more than ever before, the number of wildland firefighters was raised, as were the numbers of fire trucks and helicopters. With the SEATs in readiness, they would now try to replace the previously used large airtankers.

"Fewer than 10% of our forest fires have ever seen an airtanker," Tim Foley, Fire Manager for the western Upper Colorado River Region, asserted. Even the firefighting operations at the last large forest fires there—the Glade Park Fire in 2002, not far from the city of Grand Junction, Colorado—could not be supported by airtankers because of the prevailing wind conditions. "But it was put out all the same," said Foley.

Matt Mathes, spokesman of the U.S. Forest Service (USFS) in California, brought the concern about the safety of action forces to a point: "The airtankers now kill five percent of their pilots in every forest fire season!"

One Situation

Individual voices, to be sure, that were lost in the tumult of numerous opinions, standpoints, and utterances, but that urged a little sense in a discussion that was really directed at efforts to achieve the greatest possible safety in forest and surface firefighting. While some wanted to guarantee the greatest possible safety for the population and for the work forces by withdrawing the large airtankers, others saw the withdrawal of the airtankers as resulting in the inability to guarantee the safety of the population and the work forces anymore.

Single cases from the 2004 forest fire season could justify the latter view. In mid-July there was a forest and surface fire (the Waterfall Fire) not far from Carson City, Nevada. When the flames spread in the direction of inhabited areas and seemed to go out of control, residents looked with more and more concern at the water drops of the fire helicopters in use. "It looked as if they were using a water pistol to put out a building fire," Betty Kelly recalled. The airtankers of the Minden Aircraft Corporation—just a few miles from Carson City—could not support the 1600 firefighters at the Waterfall Fire—they were known to be "cancelled" by the USFS. One of the pilots said, "It is frustrating. We saw the need of the residents—and when my neighbor ask me why we don't take off and help, I can only regretfully shake my head." In the end, everyone in Carson City asked himself how many buildings had to burn in other forest fires before the most effective equipment—namely the large airtankers—were brought into action in the initial phase of a fire?!

Conversion—Developing and Rebuilding Aircraft for Forest and Surface Firefighting

Conversion—adapting, rebuilding, changing—that is the slogan in certain areas of firefighting when the goal is to provide specially capable, rugged firefighting vehicles as economically as possible. As subjects for rebuilding and reequipping, the handiest are military vehicles that are mustered out but still meet the criteria for special firefighting use. While conversion projects were proposed in Germany in the 1990s (for example, the firefighting tank based on the Leopard I, made by the Industrieanlagen-Betriebsgesellschaft mbH/IABG from Lichtenau, Westphalia in cooperation with the Schwing firm of Herne; the PTLF 38/150 "Water Buffalo" firefighting tank based on a Russian T-55 tank, and the MIG-21 "Hurricane" turbine-afterburner vehicle of the "System-Instandsetzungs - und Verwertungsgesellschaft" [System Repairs and Collecting Society] mbH/SIVG of Neubrandenburg, or the tracked PHF 15 T forest fire vehicle of the IBP Pietsch GmbH from Ettlingen), for decades ex-U.S. Army and Air Force vehicles have been used in the USA in the realm of forest and surface firefighting.

A successful future project in the realm of changing military aircraft into action vehicles for forest and surface firefighting is this Bell AH-1F Cobra ("Fire Snake". # N109Z) of the U.S. Forest Service (USFS). The copter was used as a lead plane.

This is true of aircraft as well as ground vehicles. Most of the American firefighting aircraft (airtankers) and some of the helicopters (such as the Bell UH-1D) come from the supplies of mustered-out Air Force aircraft. The former battle and transport planes were rebuilt and fitted with tanks and dropping apparatus for the retardants. Meanwhile, these aircraft have reached ages of over 60 years—reason enough for the U.S. Forest Service (USFS/USDA) to cancel the contracts for the use of the large airtankers in 2004 for safety reasons.

The Basis: Federal Excess Property Program (FETP)

The Federal Excess Property Personal Program (FEPP) is an initiative established by the US Congress for the further use of government vehicles by other federal organizations—particularly no longer used military vehicles, but also aircraft, personal protective equipment, technical equipment, and materials. Thus the national and state forestry and fire protection organizations rank among the beneficiaries of this program.

The initiative serves to support preventive forest fire and catastrophe protection as well as to support forest and surface firefighting in the 50 US states, and thus to be a practical completion of state and local efforts to protect lives and property in the USA.

The U.S. Forest Service (USFS), the largest super-regional forestry and forest fire agency of the U.S. Department of Agriculture (USDA), takes the role of an agent. Within the extensive U.S. Forest Service FEPP Program, the U.S, Forest Service is justified in coordinating the procurement of ex-military vehicles and passing them along to state and local firefighting organizations and fire departments in cooperation with the state forest agencies. About 70% of all vehicles and equipment thus go to county, city, and rural fire departments to support their local fire protection programs.

A significant factor in considerations for the conversion of military vehicles into forest and surface firefighting vehicles is the safety of the vehicle and its crew. Military vehicles—and this is, of course, also true of aircraft—were originally built, equipped, and used under fully different conditions and requirements than they later were to be equipped and used for in the realm of forest and surface firefighting.

General rebuilding, technical modification such as weight changes, different axle loads or centers of gravity, additional structures like tanks and pumps, plus the whole firefighting equipment, must therefore fulfill all safety-relevant requirements (Federal Motor Vehicle Safety Standards/FMVSS, National Fire Codes, National Engine Study, and in 1988 the National Wildlife Coordinating Group/NWCG).

Although conversion usually involves the further or different use of military air and ground vehicles (such as the Grumman S-2FT Tracker, Bell UH-1D or Bell AH-1 Cobra), the concept also applies to the turning of civilian vehicles into firefighting vehicles. Examples of this from the realm of aviation involve the use of civilian craft as firefighting airtankers, such as the Douglas DC-3, McDonnell Douglas C-54 Skymaster (DC-4) or Dash 8-400.

The most spectacular example of turning a large civilian plane into an airtanker is the attempt by the Evergreen firm to use a Boeing 747 as a supertanker. The first attempts and presentations with the 24,000-gallon airtanker (ca. 90,850 liters) succeeded, but the number of opponents of such future projects was large. Practitioners—instead of envisioning supertankers with tactically incalculable amounts of retardant—prefer fast, maneuverable firefighting planes with medium-sized drop volumes.

The trend toward "the more extinguishing water, the better!" at that time and resulting from the new future planning for aerial firefighting, advocated not only in the USA, appears more and more questionable!

Further, still-active conversion projects include the rebuilding of a Fairchild A-10A Thunderbolt II as a 2000-gallon airtanker (Firehog, CDF, 2006) and adapting a British Aerospace BAe 146-100 to a 3000-gallon airtanker by Minden Air in 2004-05.

Presently the world's largest airtanker? The American firm of Evergreen Aviation is testing a Boeing 747 as a supertanker. Its tanks can hold up to 24,000 gallons (ca. 90,850 liters) of water or retardant.

In California the B 747 has already been tested thoroughly, both with water (the opened hatches under the fuselage of the plane, from which the water flows, are easy to see), and also with fire retardant.

An interesting scene: Water turned to foam hits the ground. The supertanker thus lays a firebreak about 24,000 feet (about 2200 meters) long.

A Sikorsky UH-60 A Black Hawk (S-70) in its military version—at the Los Angeles County Fire Department (LACoFD) is already available as a helitanker.

Information and Tips

This book offers, in its text and photographs, a great deal of information, numerous dates, and details. What is more appropriate than to bring together the most essential data at this point in comprehensible tables, in order to provide the reader with a further fund of useful information.

The interested reader will find six tables in all:

- A comprehensive list of all known aircraft from all over the world that are or were used by fire departments, forestry or forest fire agencies, or special organizations to fight forest and surface fires.

- A summation of the most important types of aircraft used in fighting forest and surface fires, with their functions and technical data.

- A list of the airplanes and helicopters named by the United Nations (UN) and the International Search and

Rescue Advisory Group (INSARAG) for use in international major catastrophes and disasters.

- A list of the major American and Canadian manufacturers of, and charter companies that lease, firefighting aircraft.

- A list of specialized terminology for aerial firefighting and forest and surface firefighting.

In particular, Table I (international aircraft for forest and surface firefighting) offers in this form a never before published compilation of these specialized aircraft—not without gaps on a worldwide basis, but in parameters that make clear the great numbers of airtankers, fire helicopters, and jump planes.

Table I (Parts 1-13)

International Aircraft for Forest and Surface Firefighting

Listed here are the internationally known and utilized airplanes, amphibian planes, and helicopters that are or were used for forest and surface firefighting. It also lists the functions, their tactical organization by size (ICS Types, international size classification), their known area of utilization (countries), and the capacities of airtankers.

Such a list cannot be complete. The technical development in this realm is too large, the modifications in terms of the use of certain aircraft types is too manifold. Thus only the aircraft documented as being in service are listed. The listing is alphabetical, with the type designation used by the various agencies, organizations, and firms. They turn out to be widely varied—thus the presently used form of listing them must be used.

The load capacities of firefighting materials in the same types of aircraft (airtankers) sometimes vary as well. Thus the standard capacity is listed here, though it can vary according to the aircraft operators' special requirements.

The classification of their size as small, medium, and large is essentially valid internationally, though in some countries a different size and tactical classification may be chosen.

Table 2 (Parts 1-3)

Aircraft for Forest Firefighting (USA)—Technical and Tactical Data

This book is based on aerial firefighting in the USA. Thus it is appropriate to list the aircraft used there for forest and surface firefighting again with their relevant technical data. These include the length of the aircraft, the wingspan (tactically important for use at certain fire airbases), the top speed, the flight radius and the maximum range with one tank of fuel (important in long-range work), and the type of containers (fixed tanks, buckets).

Here too, standard data are given, which may be changeable depending on the technical modification of the aircraft.

Tables 3 and 4

Airplanes for Major Disasters and Catastrophes (UN-INSARAG)

To make it immediately clear, these tables include no aircraft for forest and surface firefighting! But they do offer a very interesting addition to the aircraft listed otherwise, as they see service in the most varied ways in international major disasters and catastrophes.

The lists were published by the United Nations (UN) or their International Search and Rescue Advisory Group (INSARAG). They include—in the abridged form given here—data on the speed of the aircraft, their load capacities in terms of determinable sizes and weights as well as possible numbers of passengers. In this form, the listed aircraft can also be used as transport, passenger or freight airplanes in large international forest fires.

Table 5 (Parts 1 & 2)

Manufacturers and Contractors of Aircraft for Forest Firefighting

This table lists the significant international businesses that have devoted themselves to aerial firefighting. The leading manufacturers of aircraft for forest and surface firefighting are listed—limited, though, to those who build special aircraft. General manufacturers of civilian and/or military aircraft that are used later in modified form for firefighting can be found easily in the Internet via the listed addresses.

The table states the firm's name, production site, and types of aircraft involved.

Table 6 (Parts 1 & 2)

Terminology of Aerial Firefighting and Forest and Surface Firefighting

Admittedly—sometimes it is not easy in such a specialized presentation to get by with internationally used terminology. For certain terms, mostly based on American specialized nomenclature, there is no corresponding international term—even though I have tried my best in this book. To help the interested reader understand what is written, the most important specialized terms are assembled here once again and described.

The Best-known Aircraft in International Forest and Surface Firefighting						1
Name	**Manufacturer**	**Function**	**ICS Type**	**Country**	**Tank Capacity**	
					gallons	liters
A						
A-26 »Invader«	McDonnell Douglas	Airtanker	medium	USA, Canada	1,000	3,785
A-10A »Firehog«	Fairchild-Republic, Northrop	Airtanker	II	USA	2,000	7,570
A-109	Bell/Agusta Aerospace Company	Fire Helicopter, Helitanker	medium	Italy, Switzerland, Australia, Argentina		
A-119 »Koala«	Bell/Agusta Aerospace Company	Fire Helicopter	medium	Australia, Italy		
AB-204B	Bell/Agusta Aerospace Company	Fire Helicopter	medium	Italy, Austria		
AB-205 A1	Bell/Agusta Aerospace Company	Fire Helicopter	medium	Italy		
AB-206 B-II	Bell/Agusta Aerospace Company	Fire Helicopter	medium	Italy		
AB-206 B-III	Bell/Agusta Aerospace Company	Fire Helicopter	medium	Italy		
AB-212	Bell/Agusta Aerospace Company	Fire Helicopter	medium	Austria		
AB 47G.2	Bell/Agusta Aerospace Company	Fire Helicopter	medium	Italy		
AB 47G.3 B1	Bell/Agusta Aerospace Company	Fire Helicopter	medium	Italy		
AB 412	Bell/Agusta Aerospace Company	Fire Helicopter	medium	Italy		
AB 412 CS	Bell/Agusta Aerospace Company	Fire Helicopter	medium	Italy		
AB 412 SP	Bell/Agusta Aerospace Company	Fire Helicopter	medium	Italy, Sweden		
AB 412 EP	Bell/Agusta Aerospace Company	Fire Helicopter	medium	Italy		
Aero Commander 500B	Aero Aviation Inc.	Crew- and Freight Plane, Documentation Plane, Jump Plane	medium	USA		
Aero Commander 520	Aero Aviation Inc.	Crew- and Freight Plane, Documentation Plane	medium	USA		
Aero Commander 690 »Turbo Commander«	Aero Aviation Inc.	Crew- and Freight Plane, Documentation Plane	medium	USA		
Aero Commander 840	Aero Aviation Inc.	Crew- and Freight Plane, Documentation Plane	medium	USA		
Aero Commander 1200	Aero Aviation Inc.	Crew- and Freight Plane, Documentation Plane	medium	USA		
Aeronca Champ	Aeronca Aircraft Corporation	Patrol Plane	small	USA		
Aeronca 7EC »Traveller«	Champion Aircraft Corporation	Patrol Plane	small	USA		
Aeronca 15AC »Sedan«	Aeronca Aircraft Corporation	Patrol Plane	small	USA		
Aeronca 11A »Chief«	Aeronca Aircraft Corporation	Patrol Plane	small	USA		
Aeronca 15AC »Tandem«	Aeronca Aircraft Corporation	Patrol Plane	small	USA		
AJ Savage	North American Aviation	Airtanker	II	USA		
AN-2 PP	Antonov (ASTC)	Patrol Plane, Jump Plane	medium	Russia, Poland		
AN-2 P	Antonov (ASTC)	Patrol Plane, Jump Plane, Airtanker	medium	Russia, Poland		1,200
AN-2 L	Antonov (ASTC)	Airtanker	medium	Russia, Czech Republic		2,200
AN-3T	Antonov (ASTC)	Airtanker	medium	Russia		
AN-14	Antonov (ASTC)	Airtanker, Jump Plane	large	Russia		
AN-24	Antonov (ASTC)	Jump Plane	large	Russia		
AN-26 P	Antonov (ASTC)	Airtanker, Jump Plane, Crew- and Freight Plane	large	Russia		4,000
AN-28	Antonov (ASTC)	Airtanker	large	Russia		
AN-32 P	Antonov (ASTC)	Airtanker	large	Russia, Ukraine		6,000
Ansat	Kazan Helicopters	Fire Helicopter	medium	South Korea		800, 1,000
Artic Tern	Interstate Aircraft	Patrol Plane	small	USA		
AS-350 B-1 »Astar«	Aerospatiale	Fire Helicopter, Helitanker, Helitack	III	USA		500, 800
AS-350 B-2 »Squirrel«	Aerospatiale	Fire Helicopter, Helitack	III	USA, Austria, Australia, New Zealand, France, Italy		
AS-332 »Super Puma«	Aerospatiale	Fire Helicopter	large	Germany, Switzerland, Sweden, China		
AS-341 G »Gazelle«	Aerospatiale	Fire Helicopter	small	Australia		
AS-350 B-3 »Ecureuil«	Eurocopter	Fire Helicopter	medium	Italy, France, Switzerland, Portugal, South Korea, Australia, Scotland		

The Best-known Aircraft in International Forest and Surface Firefighting						2
Name	Manufacturer	Function	ICS Type	Country	Tank Capacity	
					gallons	liters
AS-355 F-1 »TwinStar«	Aerospatiale	Fire Helicopter, Helitack	III	USA, Austria		
AS-355 F-2	Aerospatiale	Fire Helicopter	medium	Austria		
AT-802	Air Tractor Inc.	Airtanker (SEAT)	III	USA, Canada, Spain, Italy, Australia	800	3,325
AT-802A	Air Tractor Inc.	Airtanker (SEAT)	III	USA	800	3,325
AT-802AF	Air Tractor Inc.	Airtanker (SEAT)	III	USA	800	3,325
AT-802F »Fire Boss«	Air Tractor Inc.	Airtanker (SEAT)	III	USA, Portugal, Spain, Canada, Argentina	820	
AT-401B	Air Tractor Inc.	Ag Airtanker (SEAT)	small	Spain	400	1,510
AT-402A	Air Tractor Inc.	Ag Airtanker (SEAT)	small	Europe		
AT-402B	Air Tractor Inc.	Ag Airtanker (SEAT)	small	Europe		
AT-502A	Air Tractor Inc.	Airtanker (SEAT)	small	USA, Australia	500	
AT-502B	Air Tractor Inc.	Airtanker (SEAT)	small	USA	500	
AT-502	Air Tractor Inc.	Airtanker (SEAT)	small	USA, Spain, Argentina, Australia		
AT-602	Air Tractor Inc.	Airtanker (SEAT)	small	USA, Australia	630	
B						
B-234 »Chinook«	Boeing Commercial Airplanes	Fire Helicopter	large	USA		
B-727	Boeing Commercial Airplanes	Crew- and Freight Plane	large	USA		
B-737	Boeing Commercial Airplanes	Crew- and Freight Plane	large	USA		
B-747	Boeing Commercial Airplanes	Airtanker	large	USA	24,000	90,850
B-17»Flying Fortress«	Boeing Commercial Airplanes	Airtanker	large	USA		
B-17 G	Boeing Commercial Airplanes	Airtanker	large	USA		
B-24	Consolidated Vultee Aircraft Corporation	Airtanker	large	USA		
B-26	Boeing Commercial Airplanes	Airtanker	III	USA	1,200	4,540
B-26»Marauder«	Glenn L. Martin Company	Jump Plane	historic	USA (Alaska)		
B-26 B	Boeing Commercial Airplanes	Airtanker	large	Canada	1,200	4,540
B-26 C	Boeing Commercial Airplanes	Airtanker	large	Canada	1,200	4,540
Banderanti EMB-110P	Embry Air	Jump Plane		USA		
Be-12 P-200»Tschaika«	Berijew Aircraft Company	Amphibian-Airtanker	large	Russia		6,000
Be-42 / A-40»Albatros«	Berijew Aircraft Company	Amphibian-Airtanker	large	Russia		
Be-A 100	Berijew Aircraft Company	Amphibian-Airtanker	large	Russia		
Be-A 200	Berijew Aircraft Company	Amphibian-Airtanker	large	Russia, USA	2,877	12,000
Be-200 P	Berijew Aircraft Company	Amphibian-Airtanker	large	Russia	2,877	12,000
Beech»Muskateer«	Beechcraft Aircraft Corporation	Crew- and Freight Plane	small	USA		
Beech»Sport«	Beechcraft Aircraft Corporation	Crew- and Freight Plane	small	USA		
Beech»Sierra«	Beechcraft Aircraft Corporation	Crew- and Freight Plane	small	USA		
Beech»Sundowner«	Beechcraft Aircraft Corporation	Crew- and Freight Plane	small	USA		
Beech 76»Dutchess«	Beechcraft Aircraft Corporation	Crew- and Freight Plane	small	USA		
Beech 33»Bonanza«	Beechcraft Aircraft Corporation	Crew- and Freight Plane	small	USA		
Beech V35B»Bonanza«	Beechcraft Aircraft Corporation	Command Plane, Leadplane	small	USA		
Beech 36»Bonanza«	Beechcraft Aircraft Corporation	Crew- and Freight Plane	small	USA		
Beech 50»Twin Bonanza«	Beechcraft Aircraft Corporation	Crew- and Freight Plane	small	USA		
Beech 55»Baron«	Beechcraft Aircraft Corporation	Leadplane	medium	USA		
Beech 58»Baron«	Beechcraft Aircraft Corporation	Leadplane	medium	USA		
Beech 58P»Baron«	Beechcraft Aircraft Corporation	Leadplane	medium	USA		
Beech 60»Duke«	Beechcraft Aircraft Corporation	Crew- and Freight Plane	medium	USA		
Beech 65»Queen Air«	Beechcraft Aircraft Corporation	Crew- and Freight Plane	medium	USA		

The Best-known Aircraft in International Forest and Surface Firefighting						3
Name	Manufacturer	Function	ICS Type	Country	Tank Capacity	
					gallons	liters
Beech »Muskateer«	Beechcraft Aircraft Corporation	Crew- and Freight Plane	small	USA		
Beech »Sport«	Beechcraft Aircraft Corporation	Crew- and Freight Plane	small	USA		
Beech »Sierra«	Beechcraft Aircraft Corporation	Crew- and Freight Plane	small	USA		
Beech »Sundowner«	Beechcraft Aircraft Corporation	Crew- and Freight Plane	small	USA		
Beech 76 »Dutchess«	Beechcraft Aircraft Corporation	Crew- and Freight Plane	small	USA		
Beech 33 »Bonanza«	Beechcraft Aircraft Corporation	Crew- and Freight Plane	small	USA		
Beech V35B »Bonanza«	Beechcraft Aircraft Corporation	Command Plane, Leadplane	small	USA		
Beech 36 »Bonanza«	Beechcraft Aircraft Corporation	Crew- and Freight Plane	small	USA		
Beech 50 »Twin Bonanza«	Beechcraft Aircraft Corporation	Crew- and Freight Plane	small	USA		
Beech 55 »Baron«	Beechcraft Aircraft Corporation	Leadplane	medium	USA		
Beech 58 »Baron«	Beechcraft Aircraft Corporation	Leadplane	medium	USA		
Beech 58P »Baron«	Beechcraft Aircraft Corporation	Leadplane	medium	USA		
Beech 60 »Duke«	Beechcraft Aircraft Corporation	Crew- and Freight Plane	medium	USA		
Beech 65 »Queen Air«	Beechcraft Aircraft Corporation	Crew- and Freight Plane	medium	USA		
Beech A65 »Queen Air«	Beechcraft Aircraft Corporation	Crew- and Freight Plane	medium	USA		
Beech 70 »Queen Air«	Beechcraft Aircraft Corporation	Crew- and Freight Plane	medium	USA		
Beech A80 »Queen Air«	Beechcraft Aircraft Corporation	Crew- and Freight Plane	medium	USA		
Beech B80 »Queen Air«	Beechcraft Aircraft Corporation	Crew- and Freight Plane	medium	USA		
Beech B88 »Queen Air«	Beechcraft Aircraft Corporation	Crew- and Freight Plane	medium	USA		
Beech 65 »Queen Air«	Beechcraft Aircraft Corporation	Crew- and Freight Plane	medium	USA		
Beech A90 »King Air«	Beechcraft Aircraft Corporation	Crew- and Freight Plane	medium	USA		
Beech B90 »King Air«	Beechcraft Aircraft Corporation	Crew- and Freight Plane	medium	USA		
Beech C90 »King Air«	Beechcraft Aircraft Corporation	Crew- and Freight Plane	medium	USA		
Beech D90 »King Air«	Beechcraft Aircraft Corporation	Crew- and Freight Plane	medium	USA		
Beech E90 »King Air«	Beechcraft Aircraft Corporation	Crew- and Freight Plane	medium	USA		
Beech F90 »King Air«	Beechcraft Aircraft Corporation	Crew- and Freight Plane	medium	USA		
Beech B100 »King Air«	Beechcraft Aircraft Corporation	Crew- and Freight Plane	medium	USA		
Beech A100 »King Air«	Beechcraft Aircraft Corporation	Crew- and Freight Plane	medium	USA		
Beech A200 »King Air«	Beechcraft Aircraft Corporation	Jump Plane		USA		
Beech B350 »King Air«	Beechcraft Aircraft Corporation	Crew- and Freight Plane	medium	Argentina		
Beech 95 »Travel Air«	Beechcraft Aircraft Corporation	Crew- and Freight Plane, Command Plane	medium	USA, Canada		
Beech B-99 »Airliner«	Beechcraft Aircraft Corporation	Jump Plane	medium	USA		
Beech »Marquis«	Beechcraft Aircraft Corporation	Crew- and Freight Plane	medium	USA		
Beech 200 »Super King Air«	Beechcraft Aircraft Corporation	Jump Plane, Crew-Transporter	medium	USA, Australia, France		
Beech 350 »Super King Air«	Beechcraft Aircraft Corporation	Jump Plane, Crew-Transporter	medium	USA		
Bell UH-1D	Agusta/Bell Aerospace Company	Fire Helicopter	medium	Germany, Austria, Australia		
Bell UH-1F	Agusta/Bell Aerospace Company	Fire Helicopter, Helitanker, Helitack	medium	USA		
Bell UH-1H »Hueys«	Agusta/Bell Aerospace Company	Fire Helicopter, Helitanker, Helitack	medium	USA, Australia, Spain		
Bell 204B	Agusta/Bell Aerospace Company	Fire Helicopter, Helitanker, Helitack	II	USA, Australia		
Bell 204F	Agusta/Bell Aerospace Company	Fire Helicopter	medium	Australia		
Bell Super 204	Agusta/Bell Aerospace Company	Fire Helicopter, Helitanker, Helitack	II	USA		
Bell Super 205	Agusta/Bell Aerospace Company	Fire Helicopter, Helitanker, Helitack	II	USA, Australia		
Bell 205 A-1»Super Hueys«	Agusta/Bell Aerospace Company	Fire Helicopter, Helitanker, Helitack	II	USA, Australia		
Bell 206 L-4	Agusta/Bell Aerospace Company	Fire Helicopter	medium	Australia		
Bell 206 B2	Agusta/Bell Aerospace Company	Fire Helicopter	medium	Australia		
Bell 206 B3	Agusta/Bell Aerospace Company	Fire Helicopter	medium	Australia		

The Best-known Aircraft in International Forest and Surface Firefighting						4
Name	**Manufacturer**	**Function**	**ICS Type**	**Country**	**Tank Capacity**	
					gallons	liters
Bell 212	Agusta/Bell Aerospace Company	Fire Helicopter, Helitanker, Helitack	II	USA, Austria, Portugal, Spain, Canada, Australia, New Zealand		
Bell 214 B-1	Agusta/Bell Aerospace Company	Fire Helicopter, Helitanker, Helitack	II	USA, Canada, Australia, Spain		2,980
Bell 407	Agusta/Bell Aerospace Company	Fire Helicopter	III	Australia		
Bell 412	Agusta/Bell Aerospace Company	Fire Helicopter, Helitanker, Helitack	II	USA, Czech Republic, Australia, New Zealand		870
Bell 206 L-III »Long Ranger«	Agusta/Bell Aerospace Company	Fire Helicopter, Helitanker, Helitack	III	USA, South Korea, Australia		200, 400
Bell 206 L-II »Long Ranger«	Agusta/Bell Aerospace Company	Fire Helicopter, Helitanker, Helitack	III	USA, Australia		200, 400, 680
Bell 206 B-III »Jet Ranger«	Agusta/Bell Aerospace Company	Fire Helicopter, Helitanker, Helitack	III	USA, New Zealand, Australia		450
Bell AH-1»Huey Cobra«	Agusta/Bell Aerospace Company	Fire Helicopter, Helitanker	large	USA, Chile		
Bell AB-47 J »Super Ranger«	Agusta/Bell Aerospace Company	Patrol Fire Helicopter	medium	Italy		
Bell OH-58A »Kiowa«	Agusta/Bell Aerospace Company	Fire Helicopter	medium	USA, Austria		
Bell 47 »Soloy«	Agusta/Bell Aerospace Company	Fire Helicopter	medium	Australia		
BK 117	Messerschmidt-Bölkow-Blohm (MBB)	Fire Helicopter, Helitanker	medium	Australia, Greece, Croatia		1,000
BK 117	Kawasaki	Fire Helicopter, Helitanker	medium	Australia, Japan, Spain		1,000
BK 117-B2	Messerschmidt-Bölkow-Blohm (MBB)	Fire Helicopter	medium	Australia, Germany, Austria, South Korea		
BK 117 A-4	Messerschmidt-Bölkow-Blohm (MBB)	Fire Helicopter, Helitanker	III	USA, Australia		
BN-2 »Islander«	Pilatus/Britten-Norman Ltd.	Crew- and Freight Plane	medium	USA		
Bo 105	Messerschmitt-Bölkow-Blohm (MBB)	Fire Helicopter	medium	Germany, Switzerland, Australia		
Bo 105 CBS	Messerschmitt-Bölkow-Blohm (MBB)	Fire Helicopter	medium	Germany, Switzerland, Spain, Australia		
C						
C-7 »Caribou«	De Havilland Aircraft of Canada Ltd.	Jump Plane		USA		
C-27 J »Spartan«	LMATTS	MAFFS-Airtanker	large	Greece		
C-54A »Skymaster«	Fairchild Corporation	Airtanker	II	USA	2,000	7,570
C-54A »Skycrane«	Sikorsky	Helitanker	large	Australia	2,000	7,500, 9,000
C-54D »Skymaster« (DC-4)	McDonnell Douglas	Airtanker	II	USA	2,000	7,570
C-54E	McDonnell Douglas	Airtanker	II	USA	2,000	7,570
C-54F	McDonnell Douglas	Airtanker	large	Australia	2,000	7,570
C-54G	McDonnell Douglas	Airtanker	II	USA	2,200	8,320
C-54Q	McDonnell Douglas	Airtanker	large	Australia		
C-97	Boeing Commercial Airplanes	Airtanker	I	USA	3,000	11,355
C-118 »Liftmaster«	Fairchild Corporation	Airtanker	II	USA		
C-119 »Flying Boxcar«	Fairchild Corporation	Airtanker	II	USA		
C-123 K »Provider«	Fairchild Corporation	Airtanker	II	USA		
C-125 »Raider«	Northrop	Airtanker	large	USA		
C-130 »Hercules«	Lockheed Martin Corporation	Airtanker, MAFFS-Airtanker	I	USA, Germany, France	3,000	11,355
C-130A »Hercules«	Lockheed Martin Corporation	Airtanker	I	USA, France	3,000	11,355
C-212 »Aviocar«	Construcciones Aeronauticas S.A. (CASA)	Jump Plane	large	USA		

The Best-known Aircraft in International Forest and Surface Firefighting						5
Name	Manufacturer	Function	ICS Type	Country	Tank Capacity	
					gallons	liters
Cessna 120	Cessna Aircraft Company	Crew- and Freight Plane	small	USA		
Cessna 140	Cessna Aircraft Company	Crew- and Freight Plane	small	USA		
Cessna 150	Cessna Aircraft Company	Patrol Plane	small	USA		
Cessna 152	Cessna Aircraft Company	Patrol Plane	small	USA		
Cessna 170	Cessna Aircraft Company	Crew- and Freight Plane	small	USA		
Cessna 172 »Skyhawk«	Cessna Aircraft Company	Patrol Plane	small	USA, Germany		
Cessna 172 »Hawk XP«	Cessna Aircraft Company	Patrol Plane	small	USA		
Cessna 172 »Cutlass«	Cessna Aircraft Company	Patrol Plane	small	USA		
Cessna 172 »Cutlass RG«	Cessna Aircraft Company	Crew- and Freight Plane	small	USA		
Cessna 175 »Skylark«	Cessna Aircraft Company	Patrol Plane	small	USA		
Cessna 177 »Cardinal«	Cessna Aircraft Company	Patrol Plane	small	USA		
Cessna 177 »Cardinal Classic«	Cessna Aircraft Company	Patrol Plane	small	USA		
Cessna 177 »Cardinal Classic RG«	Cessna Aircraft Company	Crew- and Freight Plane	small	USA		
Cessna 180 »Skywagon«	Cessna Aircraft Company	Crew- and Freight Plane	small	USA		
Cessna 180 »Carryall«	Cessna Aircraft Company	Crew- and Freight Plane	small	USA		
Cessna 180 »Agwagon«	Cessna Aircraft Company	Crew- and Freight Plane	small	USA		
Cessna 182 »Skylane«	Cessna Aircraft Company	Patrol Plane	small	USA, Germany		
Cessna 182 »Skylane RG«	Cessna Aircraft Company	Crew- and Freight Plane	small	USA		
Cessna 185 »Skywagon«	Cessna Aircraft Company	Patrol Plane	small	USA		
Cessna 188	Cessna Aircraft Company	Patrol Plane	small	Argentina, Finland		
Cessna 205	Cessna Aircraft Company	Command Plane, Leadplane	small	USA		
Cessna 206	Cessna Aircraft Company	Command Plane, Leadplane, Jump Plane	small	USA, Argentina		
Cessna 206A	Cessna Aircraft Company	Command Plane, Leadplane	small	USA		
Cessna 206 »Stationair 6«	Cessna Aircraft Company	Command Plane, Leadplane	small	USA		
Cessna 206 »Super Skylane«	Cessna Aircraft Company	Command Plane, Leadplane	small	USA		
Cessna 207	Cessna Aircraft Company	Crew- and Freight Plane	small	USA		
Cessna 207 »Stationair 7«	Cessna Aircraft Company	Crew- and Freight Plane	small	USA		
Cessna 208	Cessna Aircraft Company	Crew- and Freight Plane	small	USA		
Cessna 208 »Caravan«	Cessna Aircraft Company	Air Attack, Jump Plane	small	France		
Cessna 208 »Grand Caravan«	Cessna Aircraft Company	Jump Plane		USA		
Cessna 208 »Stationair 8«	Cessna Aircraft Company	Air Attack	small	France		
Cessna 210 »Centurian«	Cessna Aircraft Company	Crew- and Freight Plane	small	USA, Australia		
Cessna 210 »Turbo Centurian«	Cessna Aircraft Company	Crew- and Freight Plane	small	USA		
Cessna T-303 »Crusader«	Cessna Aircraft Company	Crew- and Freight Plane	small	USA		
Cessna 310	Cessna Aircraft Company	Crew- and Freight Plane	small	USA, Europe, Australia		
Cessna 310 »Turbo«	Cessna Aircraft Company	Crew- and Freight Plane	small	USA		
Cessna 320 »Skynight«	Cessna Aircraft Company	Crew- and Freight Plane	small	USA		
Cessna 335	Cessna Aircraft Company	Crew- and Freight Plane	medium	USA		
Cessna 340	Cessna Aircraft Company	Crew- and Freight Plane	medium	USA		
Cessna 337 »Skymaster«	Cessna Aircraft Company	Command Plane, Leadplane	medium	USA, Australia		
Cessna 337 O-2	Cessna Aircraft Company	Command Plane, Leadplane	medium	USA, Australia		
Cessna 401	Cessna Aircraft Company	Crew- and Freight Plane	medium	USA		
Cessba 402 »Utiliner«	Cessna Aircraft Company	Crew- and Freight Plane	medium	USA		
Cessba 402 »Businessliner«	Cessna Aircraft Company	Crew- and Freight Plane	medium	USA		
Cessna 404	Cessna Aircraft Company	Crew- and Freight Plane	medium	Australia		
Cessna 404 »Titan«	Cessna Aircraft Company	Crew- and Freight Plane	medium	USA, Australia		
Cessna 411	Cessna Aircraft Company	Crew- and Freight Plane	medium	USA		

The Best-known Aircraft in International Forest and Surface Firefighting					6	
Name	Manufacturer	Function	ICS Type	Country	Tank Capacity	
					gallons	liters
Cessna 414	Cessna Aircraft Company	Crew- and Freight Plane	medium	USA		
Cessna 414A »Cancellor«	Cessna Aircraft Company	Crew- and Freight Plane	medium	USA, Australia		
Cessna 441 »Conquest«	Cessna Aircraft Company	Crew- and Freight Plane	medium	USA		
Cessna 421	Cessna Aircraft Company	Crew- and Freight Plane	medium	USA		
Cessna 421A	Cessna Aircraft Company	Crew- and Freight Plane	medium	USA		
Cessna 421B »Golden Eagle«	Cessna Aircraft Company	Crew- and Freight Plane	medium	USA		
Cessna 421C »Golden Eagle«	Cessna Aircraft Company	Crew- and Freight Plane	medium	USA		
Cessna 425 »Corsair«	Cessna Aircraft Company	Crew- and Freight Plane	medium	USA		
Cessna 425 »Conquest I«	Cessna Aircraft Company	Crew- and Freight Plane	medium	USA		
Cessna 441 »Conquest«	Cessna Aircraft Company	Crew- and Freight Plane	medium	USA		
Cessna 441 »Conquest II«	Cessna Aircraft Company	Crew- and Freight Plane	medium	USA		
Cessna L-19»Mountaineer«	Ector/Cessna Aircraft Company	Crew- and Freight Plane	small	USA		
CH-53 G	Sikorsky Aircraft Corporation	Fire Helicopter	large	Germany		
CH-54B »Skycrane«	Sikorsky Aircraft Corporation	Fire Helicopter	large	USA, Canada, Australia	2,500	9,460
Citabria 7GCBC	Campion/Bellanca Aircraft Corporation	Patrol Plane	small	USA		
Citabria 8GCBC »Scout«	Campion/Bellanca Aircraft Corporation	Patrol Plane	small	USA		
CL-215	Canadair Ltd./Bombardier	Amphibian-Airtanker	III	Canada, USA, Thailand, Turkey, Venezuela, Yugoslavia, Spain, Greece, Italy, Portugal, Croatia, Sweden	1,400	5,300
CL-215T	Canadair Ltd./Bombardier	Amphibian-Airtanker	III	Canada, USA, Spain	1,400	5,300
CL-415	Canadair Ltd./Bombardier	Amphibian-Airtanker	III	Canada, USA, Croatia, Greece, France, Italy	1,600	6,055
Convair 580	Consolidated Vultee Aircraft Corporation	Crew- and Freight Plane	medium	USA		
Curtiss C-46»Commando«	Curtiss-Wright	Jump Plane	historic	USA		
CV-580	Convair	Crew- and Freight Plane	large	USA, Canada, France	2,100	8,000
D						
Dash 8-Q 400 MR	De Havilland Aircraft of Canada Ltd.	Airtanker	meduim	France		10,000
DC-2	McDonnell Douglas	Jump Plane	historic	USA		
DC-3	McDonnell Douglas	Jump Plane	medium	USA		
DC-4 (C 54 »Skymaster«)	McDonnell Douglas	Airtanker	II	USA, Canada	2,000	7,570
DC-4 »Super«	McDonnell Douglas	Airtanker	II	USA	2,200	8,320
DC-6	McDonnell Douglas	Airtanker	II	USA, France	2,450	9,275
DC-6A	McDonnell Douglas	Airtanker	II	Canada	2,450	9,275
DC-6B	McDonnell Douglas	Airtanker	II	France, Canada	2,450	12,000
DC-7	McDonnell Douglas	Airtanker	I	USA	3,000	11,355
DC-7B	McDonnell Douglas	Airtanker	I	USA	3,000	11,355
DC-10	McDonnell Douglas	Airtanker	I	USA	12,000	45,420
DHC-2»Beaver«	De Havilland Aircraft of Canada Ltd.	Transport Plane, Amphibian-Airtanker, Jump Plane	IV	USA		
DHC-4»Caribou«	De Havilland Aircraft of Canada Ltd.	Amphibian-Airtanker	IV	USA		

The Best-known Aircraft in International Forest and Surface Firefighting						7
Name	Manufacturer	Function	ICS Type	Country	Tank Capacity	
					gallons	liters
DHC-6 »Twin Otter«	De Havilland Aircraft of Canada Ltd.	Jump Plane, Amphibian-Airtanker	medium	USA, Canada		
Do 228	Dornier	Jump Plane	medium	USA (Alaska)		
E						
EC 120 B »Colibri«	Eurocopter	Fire Helicopter	small	Australia		
EC 135	Eurocopter	Fire Helicopter	medium	Germany, Austria, Switzerland, Italy		
EC 145	Eurocopter	Fire Helicopter, Helitanker	medium	Australia, Switzerland, Greece, Croatia, France		1,000
EMB-110 »Bandeirante«	Embraer	Jump Plane	medium	USA		
EMB-200/2001 »Ipanema«	Embraer	Ag Airtanker	medium	Brazil		680
F						
F7F»Tigercat«	Grumman Aerospace Corporation	Airtanker	medium	USA		
F-27»Friendship«	Fokker Flugzeugwerke	Airtanker	large	USA, France	1,600	6,055
F 27-600	Fokker Flugzeugwerke	Airtanker	large	France		6,050
Fairchild »Metro«	Fairchild Corporation	Crew- and Freight Plane	large	USA		
Ford 5-AT Tri-Motor (Tin Goose)	Ford	Airtanker, Jump Plane	small	USA		
G						
G-164 »Ag-Cat«	Grumman Aerospace Corporation	Ag Airtanker	IV	USA, Greece	300	1,135
Goose G-21	Grumman Aerospace Corporation	Jump Plane	historic	USA		
G-222	Fiat u.a.	Airtanker	meduim	Italy		6,300
Gulfstream GA-7 »Cougar«	Grumman Aerospace Corporation	Crew- and Freight Plane	medium	USA		
H						
H-43»Huskie«	Kaman Aerospace Corporation	Fire Helicopter	small	USA		
H-369 C-500	Hughes Aircraft	Fire Helicopter, Patrol Helicopter	small	Italy, Australia		
H-369 HS	Hughes Aircraft	Patrol Helicopter	small	Italy		
H-500 C	Hughes Aircraft	Fire Helicopter, Patrol Helicopter	small	Australia		450
H-500 D	Hughes Aircraft	Fire Helicopter, Patrol Helicopter	small	Italy, Australia		450
HA-31»Basant«	Hindustan Aeronautics Ltd.	Ag Airtanker, Patrol Plane	small	India		
Hiller 12-D/E»Soloy«	Fairchild Corporation	Patrol Helicopter	small	USA, Australia		
Hiller FH 1100	Fairchild Corporation	Patrol Helicopter	small	USA		
HU-16B »Albatross«	Grumman Aerospace Corporation	Amphibian-Airtanker	III	USA	1,500	5,680
HU-16T	Grumman Aerospace Corporation	Amphibian-Airtanker	III	USA		
HUL-26 »Pushpak«	Hindustan Aeronautics Ltd.	Patrol Plane	small	India		
I						
IAR-818	Manicatide	Patrol Plane, Ag Airtanker	small	Romania		
IAR-822	Manicatide	Ag Airtanker	small	Romania		
Il-76 TD	Iljuschin	Airtanker	large	Russia		44,000
Il-103	Iljuschin	Ag Airtanker, Patrol Plane	small	Russia, Uzbekistan		

The Best-known Aircraft in International Forest and Surface Firefighting						8
Name	Manufacturer	Function	ICS Type	Country	Tank Capacity	
					gallons	liters
J						
JN-4D »Jenny«	Curtiss Company	Patrol Plane	small	USA		
JRM-2/3 »Mars«	Martin Aviation	Amphibian-Airtanker	I	Canada	7,200	27,250
K						
Ka-15	Kamow Helicopter	Ag Fire Helicopter	large	Russia		
Ka-26	Kamow Helicopter	Ag Fire Helicopter	large	Russia, Romania, Africa		4,000
Ka-32A	Kamow Helicopter	Fire Helicopter	large	Kanada, Spanien, Bulgarien, Griechen-land, Australien		4,000, 5,000
Ka-32A 11 BC	Kamow Helicopter	Fire Helicopter	large	Canada, Russia, Korea		
KA-32 T	Kumertau Aviation Production (Kamow)	Helitanker, Helitack	large	Russia, South Korea, Africa, Turkey		3,000, 2,000
Ka-226	Kamow Helicopter	Ag Fire Helicopter	large	Russia		
KC-97 »Stratocruiser«	Boeing Commercial Airplanes	Airtanker	I	USA (Alaska)	4,500	17,000
KH-4	Kawasaki	Fire Helicopter	large	Australia		
K-1200 K-MAX	Kaman Aerospace Corporation	Fire Helicopter	large	USA, Australia		2,500, 3,000
HH-43F »Huskie«	Kaman Aerospace Corporation	Fire Helicopter	large	USA		
L						
Lala-1	Panstwowe Zaklady Lotnicze (PZL)	Ag Airtanker	small	Poland		
L-18 »Lodestar«	Lockheed Martin Corporation	Jump Plane	historic	USA		
L-60 »Brigadyr«	Let	Ag Airtanker	small	Czech Republic, Slovakia, international	350	1,135
L-188 »Electra«	Lockheed Martin Corporation	Airtanker	II	USA, Canada	3,000	11,355
L-188 A	Lockheed Martin Corporation	Airtanker	large	Canada		
L-188 C	Lockheed Martin Corporation	Airtanker	large	Canada		
L-410 »Turbolet«	Let	Airtanker	medium	Czech Republic, Slovakia, South Korea		
Learjet 35A	Bombardier Aerospace	Crew- and Freight Plane	medium	Australia		
Let Z-37 A »Cmelak«	Zlin	Ag Airtanker	medium	Czech Republic		
Li-2 PPL	Lisunow	Patrol Plane, Jump Plane	medium	Russia		
Li-2 SCH	Lisunow	Ag Airtanker	medium	Russia		1,500
M						
M-5 »Rocket«	Maule Air Incorporation	Crew- and Freight Plane	small	USA		
M-5 »Strato Rocket«	Maule Air Incorporation	Crew- and Freight Plane	small	USA		
M5 »Lunar Rocket«	Maule Air Incorporation	Crew- and Freight Plane	small	USA		
M-20	Mooney Airplane Company (MAC)	Crew- and Freight Plane	small	USA		
M-20C »Ranger«	Mooney Airplane Company (MAC)	Crew- and Freight Plane	small	USA		
M-20D »Master«	Mooney Airplane Company (MAC)	Crew- and Freight Plane	small	USA		
M-20E »Chapparal«	Mooney Airplane Company (MAC)	Crew- and Freight Plane	small	USA		
M-20E »Super 21«	Mooney Airplane Company (MAC)	Crew- and Freight Plane	small	USA		
M-20F »Executive«	Mooney Airplane Company (MAC)	Crew- and Freight Plane	small	USA		
M-20R	Mooney Airplane Company (MAC)	Crew- and Freight Plane	small	South Korea		

The Best-known Aircraft in International Forest and Surface Firefighting					9	
Name	Manufacturer	Function	ICS Type	Country	Tank Capacity	
					gallons	liters
M-21	Mooney Airplane Company (MAC)	Crew- and Freight Plane	small	USA		
M-22 »Mustang«	Mooney Airplane Company (MAC)	Crew- and Freight Plane	small	USA		
M-201	Mooney Airplane Company (MAC)	Crew- and Freight Plane	small	USA		
M-205	Mooney Airplane Company (MAC)	Crew- and Freight Plane	small	USA		
M-252T	Mooney Airplane Company (MAC)	Crew- and Freight Plane	small	USA		
MD-500D	McDonnell Douglas	Fire Helicopter	III	USA		
MD-500E	McDonnell Douglas	Fire Helicopter	medium	Austria		
MD-530F	McDonnell Douglas	Fire Helicopter	III	USA		
MD-900 »Explorer«	McDonnell Douglas	Fire Helicopter	III	USA		
Merlin II	Swearingen Aircraft	Crew- and Freight Plane	medium	USA		
Merlin III	Swearingen Aircraft	Crew- and Freight Plane	medium	USA		
MI-2	Panstwowe Zaklady Lotnicze (PZL)	Fire Helicopter	medium	Poland		600
Mil Mi-4 P	Mil Moscow Helicopter Plant, Harbin	Ag Fire Helicopter	large	Russia, China		1,500
Mil Mi-6	Mil Moscow Helicopter Plant	Fire Helicopter	large	France		8,000
Mil Mi-8	Mil Moscow Helicopter Plant	Fire Helicopter, Helitack	large	Russia, Australia, Africa, Portugal, Spain		1,250, 3,500
Mi-8 MTV1	Mil Moscow Helicopter Plant	Fire Helicopter, Helitack	large	Russia, Australia, Africa, Slovakia, Portugal, Turkey		3,500
Mil Mi-17	Mil Moscow Helicopter Plant	Fire Helicopter	large	Slovakia		1,590
Mil Mi-26	Mil Moscow Helicopter Plant	Fire Helicopter	large	France		9,000
Mil Mi-26 TP	Mil Moscow Helicopter Plant	Fire Helicopter	large	Russia	5,200	19,680
Mil Mi-26 TC	Mil Moscow Helicopter Plant	Fire Helicopter	large	Greece		15,000
Mil Mi-171	Mil Moscow Helicopter Plant	Fire Helicopter	large	Slovakia		3,500
N						
NAS N3N	Naval Aircraft Factory	Ag Airtanker	small	USA		
NH-500D	Bredarnadi	Fire Helicopter, Patrol Helicopter	small	Italy		
Norseman	Noorduyn	Patrol Plane	small	Canada		
P						
P-2V-7 »Neptune«	Lockheed Martin Corporation	Airtanker	II	USA	2,450	9,275
P-2V-5	Lockheed Martin Corporation	Airtanker	II	USA	2,700	10,220
P3-A »Orion«	Lockheed Martin Corporation	Airtanker	I	USA	3,000	11,350
P4Y-2 »Vultee«	Consolidated Vultee Aircraft Corporation	Airtanker	II	USA		
P-61 »Black Widow«	Northrop	Airtanker	III	USA		
P-180 »Avanti«	Piaggio	Crew- and Freight Plane	medium	Italy		
PA-11»Cub«	Piper Aircraft Corporation	Patrol Plane	small	USA		
PA-12 »Super Cruiser«	Piper Aircraft Corporation	Patrol Plane	small	USA		
PA-18 »Super Cub«	Piper Aircraft Corporation	Patrol Plane	small	USA		
PA-23 »Aztec«	Piper Aircraft Corporation	Crew- and Freight Plane	small	USA		
PA-23 »Apache«	Piper Aircraft Corporation	Crew- and Freight Plane	small	USA		
PA-24 »Comanche«	Piper Aircraft Corporation	Crew- and Freight Plane	small	USA		
PA-28 »Cherokee«	Piper Aircraft Corporation	Crew- and Freight Plane	small	USA		
PA-28 »Cherokee Charger«	Piper Aircraft Corporation	Crew- and Freight Plane	small	USA		

The Best-known Aircraft in International Forest and Surface Firefighting						10
Name	Manufacturer	Function	ICS Type	Country	Tank Capacity	
					gallons	liters
PA-28 »Cherokee Flite Liner«	Piper Aircraft Corporation	Crew- and Freight Plane	small	USA		
PA-28 »Cherokee Warrior«	Piper Aircraft Corporation	Crew- and Freight Plane	small	USA, Australia		
PA-28 »Cherokee Cadet«	Piper Aircraft Corporation	Crew- and Freight Plane	small	USA		
PA-28R »Arrow IV«	Piper Aircraft Corporation	Crew- and Freight Plane	small	USA		
PA-28R »Cherokee 140«	Piper Aircraft Corporation	Crew- and Freight Plane	small	USA		
PA-28R »Arrow«	Piper Aircraft Corporation	Crew- and Freight Plane	small	USA		
PA-30 »Twin Comanche«	Piper Aircraft Corporation	Crew- and Freight Plane	small	USA		
PA-31 »Navajo«	Piper Aircraft Corporation	Crew- and Freight Plane	meduim	USA		
PA-31 »Chieftan«	Piper Aircraft Corporation	Crew- and Freight Plane	meduim	USA		
PA-31P »Mojave«	Piper Aircraft Corporation	Crew- and Freight Plane	meduim	USA		
PA-31T »Cheyenne II«	Piper Aircraft Corporation	Crew- and Freight Plane	meduim	USA		
PA-31T2 »Cheyenne II XL«	Piper Aircraft Corporation	Crew- and Freight Plane	meduim	USA		
PA-32 »Cherokee Six«	Piper Aircraft Corporation	Crew- and Freight Plane	small	USA		
PA-32R »Saratoga«	Piper Aircraft Corporation	Crew- and Freight Plane	small	USA		
PA-32R »Lance«	Piper Aircraft Corporation	Crew- and Freight Plane	small	USA		
PA-32R »Lance II«	Piper Aircraft Corporation	Crew- and Freight Plane	small	USA		
PA-34 »Seneca«	Piper Aircraft Corporation	Crew- and Freight Plane	small	USA		
PA-39 »Twin Comanche«	Piper Aircraft Corporation	Crew- and Freight Plane	small	USA		
PA-42 »Cheyenne III«	Piper Aircraft Corporation	Jump Plane, Crew- and Freight Plane	small	USA		
PA-42 »Cheyenne IIIA«	Piper Aircraft Corporation	Jump Plane, Crew- and Freight Plane, Patrol Plane	small	USA		
PA-42 »Cheyenne 400«	Piper Aircraft Corporation	Jump Plane, Crew- and Freight Plane	small	USA		
PA-44 »Seminole«	Piper Aircraft Corporation	Crew- and Freight Plane	small	USA		
PA-46 »Malibu«	Piper Aircraft Corporation	Transport Plane	small	USA		
Partenavia P-68 »Victor«	Pascale	Crew- and Freight Plane, Documentation	medium	USA		
PBY »Catalina«	Consolidated Vultee Aircraft Corporation	Amphibian-Airtanker	medium	USA		
PBY-5A Catalina	Consolidated Vultee Aircraft Corporation	Amphibian-Airtanker	medium	France		
PBY-5A »Canso«	Consolidated Vultee Aircraft Corporation	Amphibian-Airtanker	medium	Canada		
PBY-5ACF	Consolidated Vultee Aircraft Corporation	Amphibian-Airtanker	medium	Canada		
PBY-6A	Consolidated Vultee Aircraft Corporation	Amphibian-Airtanker	medium	Canada		
PB4Y2 »Privateer«	Consolidated Vultee Aircraft Corporation	Airtanker	II	USA	2,000	7,570
PBY »Super«	Consolidated Vultee Aircraft Corporation	Airtanker	III	USA	1,400	5,300
PC-6 »Porter«	Pilatus	Airtanker	medium	Switzerland, Austria		800
PC-6 B2H2 »Turbo Porter«	Pilatus	Airtanker, Jump Plane	medium	USA, Austria		800
PC-9	Pilatus	Patrol Plane	medium	Croatia		
Piper J-3 »Cub«	Piper Aircraft Corporation	Patrol Plane	small	USA		
Piper J-4 »Cub«	Piper Aircraft Corporation	Patrol Plane	small	USA		
Piper J-5 »Cub«	Piper Aircraft Corporation	Patrol Plane	small	USA		
Piper PT-1040	Piper Aircraft Corporation	Crew- and Freight Plane	small	USA		
PS-1	Shin Meiwa	Amphibian-Airtanker	large	Japan		
PV-2 »Harpoon«	Lockheed Martin Corporation	Airtanker	III	USA	800	3,330
M-15 »Belphegor«	Panstwowe Zaklady Lotnicze (PZL)	Ag Airtanker	small	Poland		

The Best-known Aircraft in International Forest and Surface Firefighting						11
Name	Manufacturer	Function	ICS Type	Country	Tank Capacity	
					gallons	liters
M-18 »Dromader«	Panstwowe Zaklady Lotnicze (PZL-Mielec)	Ag Airtanker	IV	USA, Germany, Canada, Poland, Italy, Greece, Spain, Portugal, Slovakia, Czech Republic, Australia, Cuba, Hungary, Chile, Nicaragua, Africa, China	400	1,515
M-18B »Dromader«	Panstwowe Zaklady Lotnicze (PZL-Mielec)	Ag Airtanker	IV	USA, Germany, Canada, Poland, Greece, Czechoslovakia, Australia, Cuba, Hungary, Nicaragua	400	1,515
M-18T »Dromader«	Panstwowe Zaklady Lotnicze (PZL-Mielec)	Ag Airtanker	IV	USA, Deutschland, Kanada, Poland, Greece, Czechoslovakia, Australia, Cuba, Hungary, Nicaragua	400	1,515
PZL 101 »Gawron«	Panstwowe Zaklady Lotnicze (PZL)	Ag Airtanker	small	Germany/East Germany, Poland		800
PZL 104 »Wilga«	Panstwowe Zaklady Lotnicze (PZL)	Ag Airtanker	small	Germany/East Germany, Poland		
PZL 106 A »Kruk«	Panstwowe Zaklady Lotnicze (PZL)	Ag Airtanker	small	Germany/East Germany, Poland		500
PZL 108	Panstwowe Zaklady Lotnicze (PZL)	Ag Airtanker	small	Germany/East Germany, Poland		
R						
Rockwell OV-10 »Bronco«	North American Rockwell	Leadplane	medium	USA		
Rockwell 500S »Shrike Commander«	North American Rockwell	Leadplane	medium	USA		
Rockwell 690A »Turbo Commander«	North American Rockwell	Command Plane	medium	Canada		
S						
PA-18 »Super Cub«	Piper Aircraft Corporation	Patrol Plane	small	USA		
S-2F »Firecat«	Grumman Aerospace Corporation	Airtanker	medium	France, Canada		3,295
S-2T »Turbo Firecat«	Grumman Aerospace Corporation	Airtanker	medium	France		3,295
S-2F »Tracker«	Grumman Aerospace Corporation	Airtanker	III	USA, Canada	800	3,330
S-2T »Tracker«	Grumman Aerospace Corporation	Airtanker	III	USA, France	1,200	4,540
S2R-G6 »Trush«	Ayres Corporation	Ag Airtanker (SEAT)	III	USA, Australia	200	755
S2R-T34 »Turbo Trush«	Ayres Corporation	Ag Airtanker (SEAT)	III	USA, Spain, Poland	200	755, 1,900
S2R-T45 »Turbo Trush«	Ayres Corporation	Ag Airtanker (SEAT)	III	USA, Spain, Poland, Australia	200	755
S2R-66-65 »Turbo Trush«	Ayres Corporation	Ag Airtanker (SEAT)	III	USA, Spain, Poland, Africa	200	1,500, 2,000
S-58 T	Sikorsky Aircraft Corporation	Fire Helicopter	II	USA		
SH-5	Harbin Aircraft Manufacturing Corporation	Ampibian-Airtanker	large	China		8,000
SH-60B »Seahawk«	Sikorsky Aircraft Corporation	Fire Helicopter	large	Australia		1,500
S-61 N	Sikorsky Aircraft Corporation	Fire Helicopter	I	USA, Australia		

The Best-known Aircraft in International Forest and Surface Firefighting						12
Name	Manufacturer	Function	ICS Type	Country	Tank Capacity	
					gallons	liters
S-64 »Skycrane«	Sikorsky Aircraft Corporation	Fire Helicopter	I	USA, Greece, Italy, Turkey, Canada, France, Australia, Mexico, Borneo, South Korea	2,650	up to 10,000
S-65 A	Sikorsky Aircraft Corporation	Fire Helicopter	I	USA		
S-70»Firehawk«	Sikorsky Aircraft Corporation	Fire Helicopter	large	USA	1,000	3,785
S-70A»Black Hawk«	Sikorsky Aircraft Corporation	Fire Helicopter	large	Austria		
SA-316 B »Alouette III«	Aerospatiale	Fire Helicopter	III	USA, Austria, Italy, Switzerland, Germany		
SA-315 B »Lama«	Aerospatiale	Fire Helicopter, Helitack	III	France, Switzerland, Italy		
SA 332 L »Super Puma«	Aerospatiale	Fire Helicopter	meduim	USA, Germany, Greece		
SA-342»Gazelle«	Aerospatiale/Westland	Fire Helicopter	medium	Germany		
SA-365»Dauphin II«	Aerospatiale	Fire Helicopter	medium	Japan, Italy		4,500
Sabreliner 265-80	Sabreliner Corporation	Infra-red Plane	large	USA		
Sch-2	Schawrow	Amphibian Patrol Plane	medium	Russia		
SD 330 »Sherpa«	Short Brothers PLC	Jump Plane	medium	USA		
Sherpa C-23 (Shorts 330)	Short Brothers PLC	Jump Plane	medium	USA		
Shrike Commander 500B	Aero Aviation Inc.	Crew- and Freight Plane, Documentation	medium	USA		
SP2H»Neptune«	Lockheed Martin Corporation	Airtanker	II	USA, Chile	2,000	7,570
Stearman 75 »Caydet«	Boeing Commercial Airplanes	Patrol Plane, Ag Airtanker	small	USA		
Stinson 108 »Voyager«	Stinson Aircraft Company	Patrol Plane	light	USA		
T						
Tauro 300	Anahuac	Ag Airtanker	small	Mexico		
Taylorcraft B	Taylor Brothers Airplane Company	Patrol Plane	small	USA		
Taylorcraft F-19	Taylor Brothers Airplane Company	Patrol Plane	small	USA		
Taylorcraft F-21	Taylor Brothers Airplane Company	Patrol Plane	small	USA		
Taylorcraft D	Taylor Brothers Airplane Company	Patrol Plane	small	USA		
Taylorcraft L-2	Taylor Brothers Airplane Company	Patrol Plane	small	USA		
Taylorcraft O-57	Taylor Brothers Airplane Company	Patrol Plane	small	USA		
TBM-3»Avenger«	Grumman Aerospace Corporation	Airtanker	medium	USA, Canada	600	2,270
Transall C-160	AG Transall	Airtanker	large	Germany, France		12,000
Travelair 12Q	Curtis-Wright	Jump Plane	historic	USA		
TS-2/3	Curtiss (NAF)	Ag Airtanker	small	USA		
Turbo Commander 690	Aero Aviation Inc.	Crew- and Freight Plane, Documentation	medium	USA		
Twin Beech D 18	Beechcraft Aircraft Corporation	Airtanker, Jump Plane	medium	USA		
U						
UH-60»Blackhawk«	Sikorsky Aircraft Corporation	Fire Helicopter	I	USA, Australia		1,500
UTVA-60 AG	Nikolic/Petkovic	Ag Airtanker	small	ex-Yugoslavia		
UC-64 Norseman	Noordyne Aviation Ltd	Jump Plane	historic	USA		
V						
Vertol BV-107	Boeing Commercial Airplanes	Fire Helicopter, Jump Plane	I	USA		
Vertol BV-234	Boeing Commercial Airplanes	Fire Helicopter, Jump Plane	I	USA		
Vertol CH-47»Chinook«	Boeing Commercial Airplanes	Fire Helicopter	I	USA		

The Best-known Aircraft in International Forest and Surface Firefighting						13
Name	Manufacturer	Function	ICS Type	Country	Tank Capacity	
					gallons	liters
W						
W-3»Sokol«	Panstwowe Zaklady Lotnicze (PZL)	Fire Helicopter, Patrol Helicopter	small	Poland, Czech Republic		1,600
WSK PZL-Mielec M-18T	Panstwowe Zaklady Lotnicze (PZL)	Ag Airtanker	small	USA		
Y						
Yunshuij Y-11 (C-11)	Harbin	Ag Airtanker	small	China		970
Z						
Zlin Z-37 T»Agro-Turbo«	Moravan Airplanes Inc., Zlin	Ag Airtanker	small	Czech Republic		1,000
Zlin Z-37-2»Turbo Cmelak«	Moravan Airplanes Inc., Zlin	Ag Airtanker	small	Germany/East Germany, Czech Republic		
Zlin Z-42	Moravan Airplanes Inc., Zlin	Ag Airtanker	small	Slovakia		
Zlin Z-137	Moravan Airplanes Inc., Zlin	Ag Airtanker	small	Slovakia		1,000-1,200
Z-5	Harbin Aircraft Manufacturing Corporation	Fire Helicopter	large	China		

Aircraft for Forest Firefighting (USA)—Technical & Tactical Data									1
Name	**Tactics** (purpose)	**Type** (ICS)	**Length** (feet)	**Wingspan** (feet)	**Speed** (mph)	**Radius** (feet)	**Range** (miles)	**Capacity** (gallons)	**Hatches** (number)
Beechcraft 58P »Baron«	Lead Plane		30	38	215		800		
Rockwell OV-10 »Bronco«	Lead Plane		42	40	238		1,420		
Lockheed C-130 »Hercules«	Airtanker	I	99	133	275	80	1,500	3,000	1-8
Lockheed P3-A »Orion«	Airtanker	I	116	99	290	65	1,500	3,000	8
Douglas DC-7 »Airliner«	Airtanker	I	112	128	260	81	950	3,000	6-8
Douglas DC-6 »Airliner«	Airtanker	II	111	118	240	85	1,300	2,400	6-8
Douglas DC-4 »Airliner«	Airtanker	II	93	118	220	86	1,100	2,000	4-8
Lockheed P2V »Neptune«	Airtanker	II	86	98	225	72	1,100	2,400	6
Lockheed SP2H	Airtanker	II	95	98	220	71	500	2,000	1
Consolidated PB4Y2 »Privateer«	Airtanker	II	75	110	210	68	520	2,200	6-8
Boeing KC-97	Airtanker	I	118	141	250	85	1,500	4,500	16
Canadair CL-215	Amphibian-Airtanker	III	65	94	189		550	1,300	2
Fokker F-27	Airtanker	III	83	95	276	64	500	1,600	8
Grumman S-2F »Tracker«	Airtanker	III	44	73	200	45	1,500	800	4
deHavilland DHC-2 »Beaver«	Airtanker		30	48	115	48	500	125	1
Agricultural Aircraft	SEAT	IIII	30	45	100		1,500	300–800	1
Douglas DC-3 »Turbine«	Jump Plane		58	96	210			12–20 Smokejumpers	
Shorts SD 330	Jump Plane		58	75	170			10 Smokejumpers	
deHavilland DHC-6 »Twin Otter«	Jump Plane		52	65	170			8–10 Smokejumpers	
Casa C-212	Jump Plane		50	62	195			10 Smokejumpers	
Beechcraft B-99 »Airliner«	Jump Plane		45	46	240			6 Smokejumpers	
Beechcraft 200 »Super King Air«	Jump Plane		44	55	300			6 Smokejumpers	
Dornier 228	Jump Plane		54	56	220			8 Smokejumpers	
Embraer Bandeirante	Jump Plane		48	50	200			8 Smokejumpers	
Ford Tri-Motor	Jump Plane	historic						8 Smokejumpers	
Curtiss »Travelair« 12Q	Jump Plane	historic						4 Smokejumpers	
UC-64 »Norseman«	Jump Plane	historic						4 Smokejumpers	
Douglas DC-2	Jump Plane	historic						12 Smokejumpers	
Cessna 206	Patrol-/Jump Plane	small						2 Smokejumpers	
Piper PA-42 »Cheyenne III«	Jump Plane							6 Smokejumpers	
Twin Beech D 18	Jump Plane							4 Smokejumpers	
PC-6 B2H2 »Turbo Porter«	Jump Plane							4 Smokejumpers	
Lockheed 18 »Loadstar«	Jump Plane							4 Smokejumpers	
Curtiss C-46 »Commando«	Jump Plane							32 Smokejumpers	
Martin B-26 »Marauder«	Jump Plane (Alaska)								
deHavilland DHC-2 »Beaver«	Jump Plane							4 Smokejumpers	
C-7 »Caribou«	Jump Plane							20 Smokejumpers	
Grumman »Goose« G-21	Jump Plane	historic						4 Smokejumpers	
Aero Commander 500B	Jump Plane							2 Smokejumpers	
Beechcraft A90 »King Air«	Jump Plane							6–8 Smokejumpers	
Beechcraft 200 »King Air«	Jump Plane							6–8 Smokejumpers	
US-gal = 3,785 Liter									
US-feet = 0,005 Meter									
US mph = 1,6 km/h									

Aircraft for Forest Firefighting (USA)—Technical & Tactical Data 2

Name	Tactics (purpose)	Type (ICS)	Length (feet)	Wingspan (feet)	Speed (mph)	Radius (feet)	Range (miles)	Capacity (gallons)	Hatches (number)
Banderanti EMB-110P	Jump Plane							8 Smokejumpers	
C-212 »Aviocar«	Jump Plane							8 Smokejumpers	
Sherpa C-23 (Shorts 330)	Jump Plane							12 Smokejumpers	
Dornier 228	Jump Plane							10 Smokejumpers	
Cessna 208 »Grand Caravan«	Jump Plane							6 Smokejumpers	
Piper PA-18 »Super Cub«	Patrol Plane	small	23	35	115				
Maule Rocket	Patrol Plane	small	22	30	155				
Cessna 180 »Skywagon«	Patrol Plane	small	26	36	125				
Cessna 182 »Skylane«	Patrol Plane	small	28	36	155				
Cessna 207	Transport Plane	small	32	36	155				
Cessna 210 »Centurian«	Patrol-/Transport Plane	small	28	37	190				
Cessna 337 »Skymaster«	Patrol Plane	small	30	38	170				
Cessna 210 »Centurian«	Patrol-/Transport Plane	small	28	37	190				
Cessna 337 »Skymaster«	Patrol Plane	small	30	38	170				
Piper PA-28 »Cherokee«	Transport Plane	small	23	30	120				
Piper PA-44 »Seminole«	Transport Plane	small	28	39	190				
Mooney »Ranger«	Transport Plane	small	23	35	170				
Beechcraft V35B »Bonanza«	Transport Plane	small	26	34	190				
Piper PA-23 »Aztec«	Transport Plane	small	31	37	200				
Piper PA-31 »Navajo«	Lead Plane		33	41	240				
Cessna 310	Transport Plane	small	30	38	170				
Cessna 340	Transport Plane	small	43	38	210				
Piper PA-31T »Cheyenne II«	Transport Plane	small	35	43	240				
Cessna 414A »Chancellor«	Transport Plane	small	36	44	210				
Beechcraft B80 »Queen Air«	Transport Plane	small	36	46	230				
Aero Commander 500B	Documentation Plane	small	35	45	195				
Sabreliner 265-80	Documentation Plane	small	47	45	600				
Fairchild Metro	Transport Plane	large	59	46	275				
Convair 580	Transport Plane	large	82	105	300	65			
Convair 580	Airtanker	II	82	105	300	65			
Boeing 727	Transport Plane	large	133	108	480	72			
Boeing 737	Transport Plane	large	100	93	480	58			
Bell 47 »Soloy«	Helitack	IIII	44	37	86			Bucket 96-108	
Hiller 12-D/E »Soloy«	Helitack	IIII	41	36	90			Bucket 96-108	
McDonnell Douglas MD 500D	Helitack	III	31	26	138			Bucket 96-108	
McDonnell Douglas MD 530F	Helitack	III	31	27	138			Bucket 120-144	
Bell 206 B-III »JetRanger«	Helitack	III	39	33	40			Bucket 96-108	
Bell 206 L-3 »LongRanger III«	Helitack	III	43	37	127			Bucket 96-144	
US-gal = 3,785 Liter									
US-feet = 0,005 Meter									
US mph = 1,6 km/h									

Aircraft for Forest Firefighting (USA)—Technical & Tactical Data									3
Name	**Tactics** (purpose)	**Type** (ICS)	**Length** (feet)	**Wingspan** (feet)	**Speed** (mph)	**Radius** (feet)	**Range** (miles)	**Capacity** (gallons)	**Hatches** (number)
Aerospatiale AS-350 D-1 »Astar«	Helitack	III	43	35	124			Bucket 108-144	
Aerospatiale AS-350 B-2 »Ecureuil«	Helitack	III	43	35	144			Bucket 240	
Aerospatiale AS-355 F-1 »Twin Star«	Helitack	III	43	35	132			Bucket 108-144	
Aerospatiale SA-315B »Lama«	Helitack	III	43	36	92			Bucket 180	
Aerospatiale SA-316B »Alouette III«	Helitack	III	42	36	92			Bucket 144	
MBB BO 105 CB	Rescue Helicopter	III	39	32	127			Bucket 120	
BK 117 A-4	Rescue Helicopter	III	43	36	138			Bucket 180	
Bell 204B (UH-1B)	Fire Helicopter/ Helitanker	II	55	48	104			Bucket 240	
Bell 204 »Super« (205)	Fire Helicopter/ Helitanker	II	55	48	104			Bucket 324	
Bell 205 A-1 (UH-1H)	Fire Helicopter/ Helitanker	II	57	48	104			Bucket 324	
Bell 205 »Super«	Fire Helicopter/ Helitanker	II	57	48	110			Bucket 324	
Bell 205 »Super-Duper« (Bell 212)	Fire Helicopter/ Helitanker	II	57	48	110			Bucket 324	
Bell 212 (Bell 205, UH-1N)	Fire Helicopter/ Helitanker	II	58	48	115			Bucket 324	
Bell 412	Fire Helicopter/ Helitanker	II	56	46	127			Bucket 420	
Sikorsky S-58T	Fire Helicopter/ Helitanker	II	42	56	104			Bucket 420	
Kaman H-43 »Huskie«	Fire Helicopter/ Helitanker	I	25	47	98			Bucket 324	
Bell 214 B-1	Transport Helicopter	I	62	52	160			Bucket 660-800	
Sikorsky UH-60 »Black Hawk«	Fire Helicopter/ Helitanker	I	65	54	167			Bucket 660	
Aerospatiale AS 332L »Super Puma«	Fire Helicopter	I	61	51	138			Bucket 900	
Sikorsky S-61 N (Sea King)	Fire Helicopter	I	73	62	138			Bucket 900	
Boeing Vertol 107	Fire Helicopter	I	83	50	138			Bucket 900-1,000	
Boeing 234 (CH-47 »Chinook«)	Fire Helicopter	I	99	60	155			Bucket 3000	
Sikorsky S-64 »Skycrane«	Helitanker	I	89	72	92			Tank 2000	
US-gal = 3,785 Liter									
US-feet = 0,005 Meter									
US mph = 1,6 km/h									

Airplanes for Major Disaster and Catastrophe Cases (UN-INSARAG)					
Name	**Speed** (knots)	**Load Limit** (metric tons) (2,200 lb)	**Cargo Space** L + W + H (cm)	**Cargo Hatches** W + H (cm)	**Cargo Volume** (cubic meters)
Antonov AN-12		15	1,300 x 350 x 250	310 x 240	100
Antonov AN-22		60	3,300 x 440 x 440	300 x 390	630
Antonov AN-26		5.5	1,060 x 230 x 170	200 x 160	50
Antonov AN-32		6.7	1,000 x 250 x 110	240 x 120	30
Antonov AN-72/74		10	1,000 x 210 x 220	240 x 150	45
Antonov AN-124	450	120	3,300 x 640 x 440	600 x 740	850
Airbus 300F4-100		40	3,300 x 450 x 250	360 x 260	320
Airbus 300F4-200		42	3,300 x 450 x 250	360 x 260	320
Airbus 310-200F		38	2,600 x 450 x 250	360 x 260	260
Airbus 310-300F		39	2,600 x 450 x 250	360 x 260	260
Boeing 727-100F		16	2,000 x 350 x 210	340 x 220	112
Boeing 737 200F		12	1,800 x 330 x 190	350 x 210	90
Boeing 737 300F		16	1,800 x 330 x 210	350 x 230	90
Boeing 747 100F		99	5,100 x 500 x 300	340 x 310	525
Boeing 747-200F	490	109	5,100 x 500 x 300	340 x 310	525
Boeing 747 400F		113	5,100 x 500 x 300	340 x 310	535
Boeing 757 200F		39	3,400 x 330 x 210	340 x 220	190
Boeing 767 300F		55	3,900 x 330 x 240	340 x 260	300
Douglas DC-10 10F		56	4,100 x 450 x 250	350 x 260	380
Douglas DC-10 30F		70	4,100 x 450 x 250	350 x 260	380
Ilyushin IL-76	430	40	2,500 x 330 x 340	330 x 550	180
Lockheed L-100	275	22	1,780 x 310 x 260	300 x 280	120
Lockheed L-100-20 Hercules	275	20	1,780 x 310 x 260	300 x 280	120
Lockheed L-100-30 Hercules	280	23	1,780 x 310 x 260	300 x 280	120
McDonnell-Douglas MD-11F		90	3,800 x 500 x 250	350 x 260	365

Helicopters for Major Disaster and Catastrophe Cases (UN-INSARAG)				
Name	**Speed** (knots)	**Load Limit** (kg)	**Net/Rope Load** (kg)	**Passenger Seats**
Aerospatiale SA 315B Lama	80	420	420	4
Aerospatiale SA-316B Allouette III	80	526	479	6
Aerospatiale SA 318C Allouette II	95	420	256	4
Aerospatiale AS-332L Super Puma	120	2,177	1,769	26
Bell 204B	120	599	417	11
Bell 206B-3 Jet Ranger	97	429	324	4
Bell 206L Long Ranger	110	522	431	6
Bell 412 Huey	110	862	862	13
Bell G-47	66	272	227	1
Bell 47 Soloy	75	354	318	2
Boeing H 46 Sea Knight	132			25
Boeing H 47 Chinook	130	12,210	12,210	33
Eurocopter (MBB) BO-105 CB	110	635	445	4
Eurocopter BK-117A-4	120	599	417	11
MI-8	110	3,000	3,000	20 to 30
MI-17	135	4,000	4,500	30
Sikorsky S-58T	90	1,486	1,168	12 to 18
Sikorsky S-61N	120	2,005	2,005	28
Sikorsky S-64 Skycrane	80	7,439	7,439	
Sikorsky S-70 (UH-60) Black Hawk	145	2,404	1,814	14 to 17

Manufacturers and Contractors of Aircraft for Forest Firefighting (selection)			1
Company	**Type**	**Location**	**Products**
Aero Asaki Corporation	Contractor	Tokyo, Japan	div. B-206 bis S-332, fixed wing
Aero Flight	Contractor	Kingman, Arizona	DC-4
Aero Diva	Contrractor	Costa Rica	B-205
Aero Slowakia	Contractor	Kosice, Slovakia	Zlin Z-137, C-150, C-152, C-172, Zlin Z-42, Zlin Z-37
Aero Union Corporation	Contractor	Chico, California	P2A, SP2H, C-54
Agrotors	Contractor	Gettysburg, Pennsylvania, USA	div. Bell
Airborne Fire Attack	Contractor	Aliso Viejo, California	PBY, HU-16T
Air Glacier	Contractor	Sion, Switzerland	AS-316 Alouette, AS-315 Lama, AS-350 B2
Air Tractor	Hersteller	Olney, Texas	AT-802F, AT-802, AT-602, AT-502
Air Tractor Europe	Hersteller	Sagunto, Spain	AT-802F, AT-802, AT-602, AT-502
Air Transport Europe	Contractor	Poprad, Slovakia	Mi-8
Air Spray Ltd.	Contractor	Canada	L-188, B-26 Invader, CL-215
Air Zermatt	Contractor	Zermatt, Switzerland	AS-316 Alouette, AS-315 Lama, AS-350 B2
Alpha Helicopteros	Contractor	La Reina, Chile	B-206, B-204, B-407, S-316, G-206
ARDCO, Inc.	Contractor	Tucson, Arizona	DC-4
Berliner Spezialflug GmbH	Contractor	Diepensee, Germany	Mil Mi-8T, Bell 206, Hughes, PZL
Bombardier	Hersteller	Montreal, Canada	CL-215, CL-215T, CL-415
Canadiian Air-Crane	Contractor	Delta, BC, Canada	S-64
Columbia Helicopters	Contractor	Portland, Oregon	B-234 Chinook, Vertol 107 II, AS-350B, EC-120, B-206B, B-206L, R-22, R-44
Conair Aviation	Contractor	Abbotsford, Canada	PA-60, TC-690A, Cessna 208B, DC-6, Convair 580, AT-802F, FireCat, CL-215
Copterline Oy	Contractor	Helsinki, Finland	S-76C, EC-135, BO-105, B-206, H-500
Coulson Aircrane	Contractor	Port Alberni, Canada	S-61, Bell 206
CzechAircraft s.r.o.	Hersteller	Otrokovice, Czech Republic	Z-242L, Z-143L, Z-143LSi
Downstown Aero	Contractor	Vineland, New Jersey	AgCats, Dromader
DynCorp Technical Services, Inc.	Contractor	Fort Worth, Texas	S2, S2-T, OV-10
Erickson Air Crane	Contractor	Central Point, Oregon	S-64
Evergreen	Contractor	McMinnville, Oregon	S-64 Skycrane, B-747
Evergreen Helicopters, Inc.	Contractor	McMinnville, Oregon	div. Helicopter
5-State-Helicopters	Contractor	Royse City, Texas, USA	S-58T
FarWest Helicopters	Contractor	Chilliwak, BC, Canada	B-206
Flying Firemen	Contractor	Spanaway, Washington	PBY
Flying Tankers, Inc.	Contractor	Port Alberni, Canada	Martin Mars, Cessna 210
Gateway Helicopters	Contractor	North Bay, Canada	AS-350
Hawkins & Powers	Contractor	Greybull, Wyoming	C-130A, PB4Y2, P2V
Heavy Lift Helicopters, Inc.	Contractor	Apple Valley, California	CH-54
Helicopters New Zealand	Contractor	Auckland, New Zealand	B-412, B-212, B-206, AS-350
Helicopters Victoria	Contractor	Essendon, Victoria, Australia	AS-350BA, B-206B, KingAir 350

Manufacturers and Contractors of Aircraft for Forest Firefighting (selection)			1
Company	**Type**	**Location**	**Products**
Helilagon Helicopteres	Contractor	Iceland	AS-350, AS-355
Helibravo Aviacao	Contractor	St. Domingo de Rana, Portugal	PZL W3A-2, PZL W3A-S, PZL W3A-M
Helicopteros del Sureste	Contractor	Muckamiel, Spain	B-212, B-412, B-206, A-109
Helicopteros del Sureste	Contractor	Muckamiel, Spain	B-212, B-412, B-206, A-109
HeliPortugal	Contractor	St. Domingo de Rana, Portugal	AS-350, AS-355, H-500, R-44, Kamov Ka-32
Helipro	Contractor	Palmerston North, New Zealand	R-22, MB-500, AS-350, UH-1F, BK-117,UH-12E
Helisul	Contractor	St. Domingo de Rana, Portugal	B-212
Helog AG	Contractor	Kussnacht, Switzerland	AS-350B3, Bell 407, SA-315 B Lama, Kamov Ka-32A12, AS-332 C1 Super Puma, Kaman K-1200 K-Max, Mil Mi-26
HICO Heli Kompanija	Contractor	Zagreb, Croatia	EC-130, EC-145
Hirth Air Tankers	Contractor	Buffalo, Wyoming	PV-2 Harpoon
Horizone Helicopters	Contractor	Rancho Murieta, California, USA	B-206, UH-1D
International Air Response (T&G)	Contractor	Chandler, Arizona	C-130, DC-7
Marsh Aviation	Contractor	Mesa, Arizona	S2, S2-T, THU-16, Turbo-Trush
Minden Air Corporation	Contractor	Minden, Nevada	SP2H
Neptune Aviation Services	Contractor	Alamagordo, New Mexico	P2V
Orsmond Aerial Spray Ltd/Aviation	Contractor	Bethlehem, South Africa	Ayeres S2R-T34 Trush
Ostermann Helicopter AB	Contractor	Gothenburg, Sweden	BO-105, AS-350, B-206L, B-205, EC-120
Phoenix Heliflight	Contractor	Fort McMurray, Canada	EC-120, EC-130, AS-350
Quebec Government Air Service	Contractor	Quebec, Canada	CL-215, CL-415, CL-215T
Queen Bee Air Specialities	Contractor	Rigby, Idaho	AT-802, AgCat, Trush
Rogers Helicopters	Contractor	Fresno, California, USA	B-206, B-212, B-222, UH-1B, CH-54, BO-105
Silver State Helicopters	Contractor	Las Vegas, Nevada	
Sky Bird Heli	Contractor	Kiew, Russia	Mi-8, Mi-17, Mi-26, KA-32, Il-76, div. Antonov
Skylink Aviation	Contrractor	Toronto, Ontario, Canada	Mi-26, Mi-8, B-212
SLAFCO	Contractor	Moses Lake, Washington	PBY
Talon Helicopters	Contractor	Richmond, Columbia, Canada	B-206, AS-350
TBM-Butler Aviation	Contractor	Tulare, California	C-130A, C-54, DC-6, DC-7
TransAero Helicopters	Contractor	Cheyenne, Wyoming, USA	AS-350, SA-315B, SA-316B, SA-330J
Vancouver Islands Helicopters	Contractor	Sidney, BC, Canada	B-205, B-206, H-500, R-22, B-222
Western Flying Service	Contractor	Phoenix, Arizona	
Wiskair	Contrractor	Thunder Bay, Ontario, Canada	B-206B, B-212
XL-Helicopters	Contractor	Denver, Colorado, USA	B-214 B1
Yellowhead Helicopters	Contractor	Valemont, BC, Canada	B-206B, B-206L, B-204
Moravan Airplanes Inc.	Hersteller	Zlin, Czech Republic	Z-242L, Z-42, Z-37, Z-137, u.a.

Specialist Terms of Aerial Firefighting	1
Term	**Definition**
Acre:	A measurement of surface area
Ag-Tanker:	An airplane used in agriculture
Air Attack:	Aerial support in forest and surface firefighting
Aerial Firefighting:	Fighting a fire from the air
Aerial Fire Airbase:	A base for firefighting aircraft (also Fire Airbase)
Agency:	A government forestry or forest fire department
Airtanker:	An airplane with a tank for fire-extinguishing liquid
Aircraft:	Airplanes and helicopters, flying vehicles
Attack:	A firefighting procedure
Bambi Bucket:	A flexible external water container carried by a helicopter
Base:	An airfield
Base Manager:	The director of a fire airbase
Bureau of Landmanagement (BLM):	A Federal agriculture and forestry agency
California Department of Forestry and Fire Protection (CDF):	A California state agency for forestry and firefighting, now CAL FIRE
Command Plane:	An air commander's airplane
Contractor:	A charter company that leases aircraft for firefighting
Conversion:	Rebuilding military or civilian aircraft for firefighting
Dispatch Center:	A central radio command post
Dispatcher:	A person who gives instructions by radio from a command post
Drop:	Release of extinguishing water or retardant from an aircraft
Drop Configuration:	A planned pattern or sequence of drops on a fire
Drop Zone:	The part of a fire of which drops are made
Engine:	A firefighting ground vehicle
Final Flight Pattern:	Flight to or from the scene of a fire
Fire Airbase:	An airfield for firefighting aircraft
Fire Cache:	A store of firefighting equipment and materials
Fire Crew:	A group of firefighters (Engine Crew, Helitack Crew)
Fire Helicopter:	A helicopter with one or more external containers of water or retardant
Fire Line:	A barrier created to halt the spread of a fire
Fire Retardant:	A mixture of water and chemicals to extinguish or limit a fire
Fire Trol:	A brand name of a kind of fire retardant
Fire Season:	The hot, dry part of a year
Fix(ed) Tank:	A water or retardant tank attached rigidly to an aircraft
Flanking:	A firefighting technique in which the edges of a fire are attacked
Foam:	A type of firefighting material
Forest Fire:	A fire in a woodland
Forest Firefighting:	Procedure to extinguish or control a forest fire
Gallons (gal):	A liquid measure equal to four quarts
Hand Crew:	A group of land-based firefighters
Head (of the Fire):	The foremost, hottest part of a forest or surface fire
Heavy Airtanker (Large Airtanker):	ICS Type I or II airtanker
Helibase:	An airfield for firefighting helicopters

Specialist Terms of Aerial Firefighting	2
Term	**Definition**
Helispot:	A temporary landing place for firefighting helicopters
Helitanker:	A helicopter with a fixed tank on the bottom
Helitack:	A special firefighting crew using a helicopter
Incident:	An event that requires firefighting
Incident Management System (ICS):	USA-wide fire management
Jump Plane:	A transport plane for smokejumpers
Large Airtanker:	ICS Type I or II airtanker
Leadplane / Lead Plane:	An airplane that leads firefighting aircraft to the scene of a fire
long-term (Retardant):	Extinguishing material with a long-time effect
Lookout:	A fire watchtower or observation post
MAAFS:	Modular Airborne Fire Fighting System, carried in an airplane
National Wildfire Coordinating Group (NWCG):	A connecting link for fire managers
National Interagency Fire Center (NIFC):	A central command post for forest firefighting
Phos Check:	A brand name of a kind of fire retardant
Rappeling Crew:	A specially trained fire crew that descends from a helicopter by rope
Resources:	Personnel, equipment, and materials available for forest firefighting
Retardant	A fire-extinguishing or cooling material
Retardant Line:	A strip of fire retardant dropped on a fire
Scooper:	An amphibian firefighting airplane and its apparatus for taking on water
SEAT:	Single Engine Air Tanker, a light firefighting airplane
Smokejumper:	A specially trained firefighter landed by parachute
short-term (Retardant):	Extinguishing material with a short-time effect
Surface Fire	A fire on a grassy or other non-wooded expanse
Tactics:	Methods of fighting a forest or surface fire
Type:	A size classification of firefighting aircraft
U.S. Forest Service (USFS):	A Federal forestry and forest firefighting agency
Watertender:	A ground vehicle that supplies firefighting water
Wildfire (Wildland Fire):	A fire in an inaccessible region

Bibliography

»Aircraft Rescue and Fire Fighting«, Ausgabe 3, International Fire Service Training Association, Oklahoma State University, 1992; ISBN 0-87939-099-9

»Aircraft Use Report«, USDI Office of Aircraft Services (OAS; 23 9/91)

»Aircraft Identification Guide«, National Wildfire Coordinating Group (NWCG), Boise/Idaho, 1994 (updated)

»Interagency Airtanker Base Planning Guide«, Ausgabe 3, National Wildfire Coordinating Group (NWCG), Fire Equipment Working Team, 1995; National Interagency Fire Center (NIFC), Boise/Idaho

»Interagency Helicopter Operations Guide«, 1998, National Interagency Fire Center (NIFC), Boise/Idaho

»Aviation Mishap Information System (AMIS) Incident/ Aviation Hazard/Maintenance Deficiency Report«, Office of Aircraft Services (OAS/USDI; 34 3/92), Boise/ Idaho

»Basic Aviation Safety«, Office of Aircraft Services (OAS/USDI), Boise/Idaho, 1991

»CDF Airtankers Test Dropping New Colored Retardants«, News Release 27. August 2005, California Department of Forestry and Fire Protection (CDF), Sacramento/California

Clayton, Bill, Day, David und McFadden, Jim: »Wildland Firefighting«, State of California, Resources Agency, Department of Forestry and Fire Protection, 1986

»Contract Termination for Large Airtankers«, National Interagency Fire Center (NIFC), Press Release 10. May 2004, Boise/Idaho

»Cumulative Aircraft Use/Payment Summary«, US Forest Service (USFS/USDA; 6300-49 3/94)

»Federal Aerial Firefighting: Assessing Safety and Effectiveness, Blue Ribbon Panel«, Report U.S. Forest Service (USFS/USDA), US Bureau of Land Management (BLM/USDI), Dezember 2002

»Firefighters Guide«, National Wildfire Coordinating Group (NWCG), Boise/Idaho, 1986

»Flight Use Report». US Forest Service (USFS/USDA; 6500-122 8/95)

Fuller, Margaret: »Forest Fires – An Introduction to Wildland Fire Behavier, Management, Firefighting and Prevention«, Verlag Wiley Nature Editions, 1991, ISBN 0-471-52189-2

Hall, Stephen R.: »Consolidation and Analysis of Loading Data in Firefighting Operations – Analysis of Existing Data and Definition of Preliminary Air Tanker and Lead Aircraft Spectra«, Abschlussbericht Oktober 2005, U.S. Department of Transportation, Federal Aviation Administration; National Technical Information Service (NTIS), Springfield/Virginia

Hall, Stephen R., »The Impact of Low-Level Roles on Aircraft Structural Integrity with particular Reference to Firebombers«, National Research Council of Canada, Contractor Report CR-SMPL-2002-0258, November 2002

Hall, Stephen R., »A Test Specifcation to Quantify the Load Response of a Large Aircraft involved in Aerial Firefighting Operations«, Celeris Aerospace Canada Inc., Technical Report CAC/TR/02-005 Revision A, Februar 2003

Hall, Stephen R., »Activities Pertaining to Aerial Firefighting Aircraft Undertaken During the 2003 Aerial Firefighting Season«, Presentation to the FAA Air Tanker Working Group (ATWG), Washington/D.C., Dezember 2003

Hall, S.R., Perry, R. und Braun, J.F., »The Safe and Economic Structural Health Management of Air Tanker and Lead Aircraft involved in Firebombing Operations«, Vortrag anlässlich des »7. Annual Wildland Fire Safety Summit«, Toronto/Ontario, Canada, November 2003

»Initial Report of Aircraft Mishap«, U.S. Office of Aircraft Services (OAS/USDI), Boise/Idaho, 1993

»Interagency Contract Information for Airtanker, Helicopter, Large Transport, and Smokejumper Aircraft«, U.S. Forest Service (USFS/USDA), National Contracting Office, Boise/Idaho

»Interagency Helicopter Training Guide« (S-217), National Wildfire Coordination Group (NWCG), Student Workbook, 1993

»Lot Acceptance, Quality Assurance, and Field Quality Control for Fire Retardant Chemicals«, National Wildfire Coordination Group (NWCG), Fire Equipment Working Team, 1995

»Interagency Airtanker Base Directory« (IABD), National Interagency Fire Center (NIFC), Boise/Idaho

»Interagency Airspace Coordination Guide« (IACG), Interagency Airspace Committee, Office of Aircraft Services (OAS/USDI), Boise/Idaho

Jendsch, Wolfgang: »Firefighters and Fire Trucks – Feuerwehren in den USA«, Fachbuch »BRAND – Die

Feuerwehren der Welt«, Chronik 5, 1997, Weltrundschau Verlag, Baar/Schweiz

Jendsch, Wolfgang: »In Deserts and Wildlands – Wildland Firefighting in den USA«, Fachbuch »Brand – Die Feuerwehren der Welt«, Chronik 6, 1998, Weltrundschau Verlag, Baar/Schweiz
Jendsch, Wolfgang: »Go West – Bei den Feuerwehren in Idaho und Kalifornien«, »Feuerwehr-Magazin«, Ausgabe 7/1999, Kortlepel Verlag

Jendsch, Wolfgang: »Löschflieger«, Fachbuch »BRAND – Die Feuerwehren der Welt«, Chronik 7, 1999, Weltrundschau Verlag, Baar/Schweiz

Jendsch, Wolfgang: »California High – Kern County Fire Department«, Fachbuch »BRAND – Die Feuerwehren der Welt«, Chronik 7, 1999, Weltrundschau Verlag, Baar/ Schweiz

Jendsch, Wolfgang: »25 Years after the Wildfire Disaster in Niedersachsen – Modifications of Fire Protection Technique", IAWF »Wildfire Magazine« (USA), Ausgabe Februar/März 2000

Jendsch, Wolfgang: »Let's go to the Fire! – Wald- und Flächenbrandbekämpfung in den USA«, Fachzeitschrift »Brandschutz – Deutsche Feuerwehr-Zeitung«, Ausgabe 7/2000, Kohlhammer Verlag

Jendsch, Wolfgang: »Cerro Grande Fire – Waldbrandbekämpfung bei Los Alamos, New Mexiko/USA«, Fachzeitschrift »Brandschutz – Deutsche Feuerwehr-Zeitung«, Ausgabe 7/2000, Kohlhammer Verlag

Jendsch, Wolfgang: »Burning Disaster – Wald- und Flächenbrände in den Weststaaten der USA«, »Feuerwehr-Magazin«, Ausgabe 12/2000, Kortlepel Verlag

Jendsch, Wolfgang: »Durch Gras, Wüsten und Brushlands – Waldbrandlöschfahrzeuge in den USA«, Fachbuch »BRAND – Die Feuerwehren der Welt«, Chronik 8/2000, Weltrundschau Verlag, Baar/Schweiz

Jendsch, Wolfgang: »Planung für das Disaster«, Fachzeitschrift »Brandschutz – Deutsche Feuerwehr-Zeitung«, Ausgabe 04/2001, Kohlhammer Verlag

Jendsch, Wolfgang: »Helitack – Hubschrauber im Waldbrandeinsatz«, Fachbuch »BRAND – Die Feuerwehren der Welt«, Band 2001, Weltrundschau Verlag, Baar/Schweiz

Jendsch, Wolfgang: »Bartons Erbe – Air Attack des Los Angeles County Fire Department", Fachbuch »BRAND – Die Feuerwehren der Welt«, Band 2001, Weltrundschau Verlag, Baar/Schweiz

Jendsch, Wolfgang: »The Big Ones – North Tree Fire (NTF), USA", »Feuerwehr-Magazin«, Ausgabe 12/2001, Kortlepel Verlag

Jendsch, Wolfgang: »North Tree Fire (NTF) – Kommerzieller Brandschutz aus Monterey«, Fachbuch »BRAND – Die Feuerwehren der Welt«, Band 2002, Weltrundschau Verlag, Baar/Schweiz

Jendsch, Wolfgang: »Prepared for the Fires – Wald- und Flächenbrandbekämpfung in den USA«, Fachmagazin »Brandaus«, LFV Niederösterreich, Ausgabe Mai 2003

Jendsch, Wolfgang: »T-130 down! – Absturz des Löschflugzeuges 130«, Fachbuch »BRAND – Die Feuerwehren der Welt«, Band 2003, Weltrundschau Verlag, Baar/ Schweiz

Jendsch, Wolfgang: »Waldbrände in Europe – Michaela: Hot and dangerous!", Fachbuch »BRAND – Die Feuerwehren der Welt«, Band 2003, Weltrundschau Verlag, Baar/Schweiz

Jendsch, Wolfgang: »Brandursache: Teufelswinde! – Die Waldbrandkatastrophe 2003 in Kalifornien«, Fachzeitschrift »Brandschutz – Deutsche Feuerwehr-Zeitung«, Ausgabe 1/2004, Kohlhammer Verlag

Jendsch, Wolfgang: »Das Ende der Heavy Airtanker?«, Fachbuch »BRAND – Die Feuerwehren der Welt«, Band 2004, Weltrundschau Verlag, Baar/Schweiz

Jendsch, Wolfgang: »Effektiv und bewährt: Das amerikanische Incident Command System (ICS)", Fachbuch »BRAND – Die Feuerwehren der Welt«, Band 2004, Weltrundschau Verlag, Baar/Schweiz

Jewel Jr., J.W., Morris, G.J., und Avery, D.E., »Operating Experiences of Retardant Bombers during Firefighting Operations«, NASA-TM-X-72622, November 1974

Kopenhagen, Wilfried und Beeck, Jochen K.: »Das große Flugzeu-Typenbuch«, Motorbuch Verlag, 2005, ISBN
3-613-02522-1

Linkewich, A.: »Air Attack on Forest Fires: History and Techniques«, ISBN 0-09690262-0-X, Verlag D.W. Friesen and Sons, Calgary/Alberta, 1972

»Lot Acceptance, Quality Assurance, and Field Quality Control for Fire Retardant Chemicals«, 6. Auflage Mai 2000, National Wildfire Coordinanting Group (USDA, USDI)

Lowe, Joseph D.: »Wildland Firefighting Practices«, Verlag Delmar/Thomson, 2000, ISBN 0-7668-0147-0

»National Interagency Mobilization Guide« (NIMG), National Interagency Fire Center (NIFC), National Incident Coordination Center (NICC), Boise/Idaho

»National Long Term Fire Retardant Requirements Contract«, US Forest Service (USFS/USDA), National Contracting Office, Boise/Idaho

Parfitt, M.C. und Hall, Stephen R.: »A Preliminary Comparison of Aerial Firefighting and Anti-Submarine Warfare (ASW) Aircraft Load Spectra«, Celeris Aerospace Canada Inc., Technical Report CAC/TR/03-002, 23 May 2003

»Publications – National Fire Equipment System Catalog Part 2«, National Wildfire Coordinating Group (NWCG), Boise/Idaho, 1999

Smith, Barry D.: »Fire Bombers in Action«, Motorbooks International, ISBN 0-7603-0043-7, 1995

Teie, William C.: »Study Guide – Firefighters Handbook on Wildland Firefighting«, Verlag Deer Valley Press, Rescue/California, 2003, ISBN 1-931301-15-8

»USDA/USDI Aircraft Radio Communications and Frequency Guide«, US Forest Service (USFS/USDA), National Incident Radio Support Cache, Avionics Section, 1989; National Interagency Fire Center (NIFC), Boise/Idaho

»USDA Forest Service Manual 5700«, US Forest Service (USFS/USDA), National Aviation Operations, National Interagency Fire Center (NIFC), Boise/Idaho

»USDI Departmental Manual Aviation Management« (350-354, einschließlich »Aviation Fuel Handling Handbook»), Office of Aircraft Services (OAS/USDI), Boise/Idaho

»Wildland Fire Supression Tactics Reference Guide«, National Wildfire Coordinating Group (NWCG), Boise/Idaho, 1996

»Standard Operational Procedures Handbook – SEAT«, Bureau of Landmanagement (BLM/USDI), Office of Fire and Aviation, National Interagency Fire Center (NIFC), Boise/Idaho

»Interagency Single Engine Air Tanker Operations Guide« (ISOG), National Aviation Operations, National Interagency Fire Center (NIFC), Boise/Idaho

Photo Credits

Umschlagbilder vorn: Wolfgang Jendsch, 02/03 Frank Garcia, 04/05 Jendsch, 05-1 Archiv Jendsch, 06/07 Jendsch, 08/09 Jendsch, 14-1 Mike Lynn, 14-2 Mike Lynn, 15-1 Mike Lynn, 15-2 Mike Lynn, 16-1 Mike Lynn, 17-1 Mike Lynn, 17-2 Mike Lynn, 19-1 Cedric Soriano, 21-1 Sinisa Jembrih, 22/23 Jendsch, 26-1 Jendsch, 27-1 Jendsch, 29-1 Frank Garcia, 30-1 Jendsch, 31-1 Jendsch, 31-2 Jendsch, 33-1 Mike Lynn, 33-2 Jendsch, 34-1 Jendsch, 35-1 Jendsch, 35-2 Jendsch, 36-1 Jendsch, 36-2 Jendsch, 36-3 Jendsch, 37-1 Jendsch, 38/39 Jendsch, 40-1 Jendsch, 40-2 Jendsch, 41-1 Pat McKelvey, 42-1 Jendsch, 43-1 Jendsch, 44-1 Roman Kotelnikov/Avialesookhrana, 44-2 Roman Kotelnikov/Avialesookhrana, 45-1 Jendsch, 45-2 Jendsch, 45-3 Jendsch, 45-4 Jendsch, 46-1 Michael Schnaufer, 47-1 NATO Media Library, 47-2 Jiri Sucharda, 48-1 Jendsch, 48-2 Jiri Sucharda, 49-1 Jiri Sucharda, 49-2 Jendsch, 49-3 Jendsch, 50-1 Jendsch, 51-1 Cedric Soriano, 51-2 Cedric Soriano, 51-3 Cedric Soriano, 52-1 Jendsch, 53-1 Jendsch, 53-2 Jendsch, 54-1 Jendsch, 54-2 Jendsch, 54-3 Jendsch, 55-1 Jendsch, 55-2 Jendsch, 55-3 Jendsch, 56-1 Jendsch, 56-2 Jendsch, 56-3 Jendsch, 57-1 Jendsch, 57-2 Jendsch, 57-3 Jendsch, 57-4 Jendsch, 58-1 Jendsch, 59-1 Jendsch, 59-2 Jendsch, 59-3 Jendsch, 60-1 Jendsch, 60-2 Jendsch, 60-3 Jendsch, 61-1 Jendsch, 61-2 Jendsch, 61-3 Jendsch, 62-1 Jendsch, 62-2 Jendsch, 63-1 Jendsch, 63-2 Jendsch, 63-3 Dimitris Fragkias, 65-1 Jendsch, 67-1 Jendsch, 67-2 Jendsch, 68-1 Jendsch, 68-2 Jendsch, 68/69 Jendsch, 71-1 Jendsch, 71-2 Jendsch, 72/73 Jendsch, 73-1 Jendsch, 74-1 Mike Lynn, 74-2 Mike Lynn, 74-3 Mike Lynn, 74/75 Jendsch, 76/77 Jendsch, 78-1 Jendsch, 78-2 Jendsch, 78-3 Jendsch, 79-1 Jendsch, 79-2 Jendsch, 80/81 Jendsch, 80-2 Jendsch, 81-2 Archiv NIFC, 82-1 Jendsch, 82-2 Jendsch, 82-3 Jendsch, 83-1 Jendsch, 83-2 Jendsch, 83-3 Jendsch, 84-1 Jendsch, 84-2 Jendsch, 84-3 Jendsch, 85-1 Jendsch, 85-2 Jendsch, 85-3 Mike Lynn, 86/87 Jendsch, 87-1 Mike Lynn, 87-2 Mike Lynn, 87-3/88-2 Jendsch, 88-1 Jendsch, 90-1 Mike Lynn, 90-2 Mike Lynn, 91-1 Jendsch, 91-2 Jendsch, 92/93 Jendsch, 93-1 Jendsch, 93-2 Jendsch, 94-1 Mike Lynn, 94-2 Mike Lynn, 95-1 Jendsch, 95-2 Jendsch, 96/97 Jendsch, 98-1 Jendsch, 98-2 Jendsch, 99-1 Jendsch, 100-1 Jendsch, 101-1 Jendsch, 101-2 Jendsch, 102-1 Jendsch, 102-2 Jendsch, 103-1 Jendsch, 103-2 Jendsch, 104-1 Mike Lynn, 104-2 Mike Lynn, 105-1 Mike Lynn, 104-3/105-2 Mike Lynn, 106/107 Jendsch, 108-1 Frank Garcia, 108-2 CDF, 109-1 Mike Lynn, 109-2 Jendsch, 110/111 Jendsch, 112/113 Jendsch, 113-1 Jendsch, 114-1 Jendsch, 115-1 Jendsch, 115-2 Conair/Archiv

Jendsch, 116-1 Archiv Air Tractor, 117-1 Air Tractor, 117-2 Mike Lynn, 118-1 Ivo Mitachek, 118-2 Ivo Mitacek, 119-1 Archiv Jendsch, 119-2 Jendsch, 121-1 Air Tractor, 122-1 Air Tractor, 123-1 Air Tractor, 124-1 Dimitris Fragkias, 124-2 Jendsch,, 125-1 Dimitris Fragkias, 126-1 Ivo Mitacek, 126-2 Dimitris Fragkias, 127-1 Roman Vana, 127-2 Ivo Mitacek, 128-1 Air Tractor, 128-2/129-2 Jendsch, 129-1 Air Tractor, 129-3 Air Tractor, 130/131 Air Tractor, 132 Archiv Jendsch, 136/137 Jendsch, 139-1 Jendsch, 139-2 Pat McKelvey, 142-1 Pete Dobbins, 142-2 Pat McKelvey, 143-1 Jendsch, 143-2 Jendsch, 144-1 Jendsch, 144-2 Jendsch, 145-1 Jendsch, 145-2 Jendsch, 146-1 Jendsch, 147-1 Jendsch, 148/149 Jendsch, 150-1 Jendsch, 150-2 Jendsch, 150-3 Jendsch, 150-4 Jendsch, 151-1 Roland Oster, 151-2 Roland Oster, 152-1 Frank Garcia, 152-2 Jendsch, 153-1 Roland Oster, 154-1 Jendsch, 154-2 Roman Kotelnikov/Avialesookhrana, 155-1 Jendsch, 155-2 Jendsch, 157-1 David Thomas, 158-1 David Thomas, 160-1 Erickson Aircrane, 161-1 David Thomas, 161-2 Silver State Helicopters, North Las Vegas, NV/USA, 162-1 Woody Chain, 162-2 Erickson Aircrane, 163-1 Jendsch, 163-2 Jendsch, 163-3 Jendsch, 164/165 Jendsch, 165-2 Jendsch, 166/167 Jendsch, 166-2 Jendsch, 167-2 Jendsch, 169-1 Jendsch, 169-2 Jendsch, 170/171 NIFC/ Archiv Jendsch, 172-1 Roman Kotelnikov/Avialesookhrana, 173-1 NIFC/Archiv Jendsch, 173-2 Woody Chain, 174-1 Jendsch, 174-2 Archiv Jendsch, 175-1 Jendsch, 175-2 Jendsch, 175-3 Jendsch, 175-4 Jendsch, 176-1 Jendsch, 176-2 Jendsch, 177-1 Jendsch, 177-2 Jendsch, 177-3 Jendsch, 178/179 Jendsch, 180-1 Jendsch, 180-2 Jendsch, 181-1 Jendsch, 181-2 Jendsch, 182/183 Jendsch, 184/185 Jendsch, 186-1 Jendsch, 186-2 Jendsch, 187-1 Jendsch, 187-2 Jendsch, 188-1 Jendsch, 188-2/189-2 Jendsch, 189-1 Jendsch, 190-1 Jendsch, 192-1 Jendsch, 192-2 Jendsch, 193-1 Jendsch, 194-1 Jendsch, 194-2 Jendsch, 195-1 Mike Lynn, 195-2 Archiv Mike Lynn, 195-3 Mike Lynn, 195-4 Mike Lynn, 196/197 Jendsch, 198/199 Roman Kotelnikov/Avialesookhrana, 199-2 Roman Kotelnikov/ Avialesookhrana, 200-1 Jendsch, 200-2 Jendsch, 201-1 Jendsch, 201-2 Jendsch, 202-1 Jendsch, 203-1 Jendsch, 203-2 Jendsch, 204/205 Jendsch, 204-2 Jendsch, 205-2 Jendsch, 206-1 Jendsch, 206-2 Jendsch, 207-1 Jendsch, 208-1 Jendsch, 208-2 Jendsch, 209-1 Jendsch, 209-2 Jendsch, 210/211 Cedric Soriano, 213-1 Benoit Tenis, 214-1 Cedric Soriano, 215-1 Roman Vana, 215-2 Sinisa Jembrih, 216-1 Roman Kotelnikov/ Avialesookhrana, 216-2 Air Tractor, 216-3 Jendsch, 218-1 Roland Oster, 218-2/ 218-1 Bundespolizei, 219-1 Jendsch, 219-2 Jendsch, 220-1 Jendsch, 220-2 Jendsch, 220-3/221 Jendsch, 222/223-1 HELOG/Archiv Oster, 223-2 Jürgen Fischer/Heer, 224-1/225-1 Roland Oster, 224-2 Toni Dahmen/ Luftwaffe, 225-2 Jendsch, 227/228 Bundespolizei, 228-1 Jendsch, 228-2 Jendsch, 229-1 Jendsch, 229-2 Jendsch, 230-1 Jendsch, 231-1 Jendsch, 232-1 Rega Schweiz, 233-1 Roland Oster, 234/235 Jendsch, 236-1 Jendsch, 237-1 Cedric Soriano, 238-1 Cedric Soriano, 239-1 Cedric Soriano,, 239-2 Cedric Soriano, 240-1 Michael Schnaufer, 240-2 Michael Schnaufer, 241-1 Cedric Soriano, 242/243 Dimitris Fragkias, 244-1 Jendsch, 245-1 Dimitris Fragkias, 246/247 Dimitris Fragkias, 248-1 Dimitris Fragkias, 248-2/249-2 Dimitris Fragkias, 249-1 Agusta Aerospace Corporation, 250-1 NATO, 250-2 Jendsch, 251-1 Agusta Aerospace Corporation, 251-2 Jendsch, 252/253-1 Jendsch, 253-2 Jendsch, 254-1 Sinisa Jembrih, 254-2/255 Sinisa Jembrih, 256-1 Roman Vana, 256-2 Ivo Mitacek, 257-1 Roman Vana, 257-2 Roman Vana, 258/259 Jendsch, 260-1/261 Jendsch, 260-2/261 Jendsch, 262/263-1 Roman Vana, 263-1 Robert Kozia, 263-2 Robert Kozia, 263-3 Tomasz Nieslony, 264-1 Roman Kotelnikov/Avialesookhrana, 265-1 Roman Kotelnikov/ Avialesookhrana, 266-1 Anthony Gray, 267-1 Jake Oosthuizen, 269-1 Staatliche Forstbehörde Südkorea, 270-1 Roman Kotelnikov/Avialesookhrana, 270-2/271-2 Archiv Jendsch, 271-1 Roland Oster, 272/273 Roman Kotelnikov/Avialesookhrana, 274-1 Staatliche Forstbehörde Südkorea, 274-2 Roland Oster, 275-1 Eurocopter, 275-2 Cedric Soriano, 276-1 Richard Jud, 277-1 Richard Jud, 277-2 Richard Jud, 278-1 Staatliche Forstbehörde Südkorea, 278-2 Staatliche Forstbehörde Südkorea, 279-1 Staatliche Forstbehörde Südkorea, 279-2 Staatliche Forstbehörde Südkorea, 281-1 Dimitris Fragkias, 282-1 Spanische Luftwaffe Ejercito del Aire, Archiv Michael Hase, 283-1 Spanische Luftwaffe Ejercito del Aire, Archiv Michael Hase, 283-2 Jendsch, 284-1 Sinisa Jembrih, 284-2 Michael Hase, 285-1 Cedric Soriano, 285-2 Jendsch, 284-3/285-3 Dimitris Fragkias, 286-1 Cedric Soriano, 287-1 Jendsch, 287-2 Air Tractor, 288/289 Frank Garcia, 290-1 Frank Garcia, 293-1 Air Tractor, ,295-1 Jendsch, 296/297 Jendsch, 299-1 Jendsch/Focus TV, 299-2 Jendsch/ Focus TV, 299-3 Jendsch/Focus TV, 299-4 Jendsch/ Focus TV, 300/301-2 Jendsch, 301-1 Jendsch, 302/303-1 Jendsch, 303-2 Jendsch, 304/305 Jendsch, 306-1 Jendsch, 307-1 Jendsch, 308/309 Jendsch, 316-1 Mike Lynn, 318-1 Evergreen Aviation, 318-2 Evergreen Aviation, 319-1 Evergreen Aviation, 320-1 Evergreen Aviation, 320-2/321 Roland Oster, 322-1 Jendsch,
324-1 Jendsch, 337-1 Jendsch, 346-1 Archiv Jendsch.

Tables: Wolfgang Jendsch
Front cover picture: Bombardier Inc.
Back cover picture: Cedric Soriano, Wolfgang Jendsch

Author's Note

This book includes countless technical and informative descriptions, tips, and technical data. I have deliberately chosen and formulated them so as to make them understandable to non-specialist readers in the realm of firefighting and air traffic. Despite thorough and careful efforts, this does not rule out the possibility that some of the material on the subject of flight technology may be imprecise or even faulty. In addition in terms of certain technical information I may have been referred to information that is contradictory or vocally misunderstood.

In such cases I ask for understanding and patience. I will gladly accept completing or correcting information, which I can then include in a possible new edition of this book.

The pictures published here come largely from my own archives of more than forty thousand photos, as well as from authors and photographers from all over the world. I have taken care to arrange and identify the pictures correctly. Possible errors were not intentional, and here too, I ask you kindly for understanding.

Wolfgang Jendsch
info@feuerwehrprese.de

Hearty Thanks to All Collaborators

A book like this really cannot be written alone—certainly not when it is to be accurate on an international basis. Based as it is on my own very extensive information and photographs, as well as on personal specialist experiences, it still required numerous collaborators who, in the preliminaries as well as the actual compiling of the book, were active in providing assistance, acquiring information, obtaining photographs or "opening doors" to the organizations and institutions that would otherwise have been scarcely or not at all accessible to me.

For all of this indispensable and valuable support, I hereby offer my heartiest thanks. My thanks go out equally to all those who helped, even if I can only name certain persons and organizations as examples and representatives of all the many others.

First of all, I offer hearty thanks to all the friends and firefighting comrades from the western states of the USA who stood beside me over the years on my visits to numerous fire airbases and smokejumper bases in Washington, Oregon, Idaho, Montana, California, Nevada, Utah, Wyoming, Colorado, Arizona, and New Mexico—especially to the Battalion Chief and Operation Section Chief of the California Interagency Incident Management Team #1, Geoff Wilford, who always encouraged my activity in the USA and supported it with valuable information.

In addition, I would like to thank the many fire managers of the various facilities and organizations, particularly Forest Aviation Officer Jim Boukidis and Air Base Manager

Patrick Basek of the Fresno Air Attack Base in California, Base Manager Ed Ish of the Porterville Airtanker Base, Base Managers Dennis Gregorkiewicz and Jo Lopour of the Cedar City Airtanker and Smokejumper Base in Utah, Education Foreman Wayne Williams and Training Foreman Everett K. Weniger of the USFS Smokejumper Base in Missoula, Montana, the late Air Service Manager C. W. "Bill" Parks of the Libby Airtanker Base (USFS) in Fort Huachuca, Arizona, Base Manager/COR Rance Irwin of the Airtanker and Smokejumper Base (USFS) in Silver City, New Mexico, Fire Investigator Bruce Moran of the U.S. Air Force Base at Indian Springs, Nevada, and Public Information Officer (PIO) Janelle Smith of the National Interagency Fire Center (NIFC) in Idaho, Karen Terill of the California Department of Forestry and Fire Protection (CDF) Headquarters in Sacramento, and Chuck Dickson of the Kern County Fire Department (KCFD) in California.

Equally hearty thanks go out—here too, these are representative of all those not named here—to airtanker pilot Pat Leroux, his co-pilot Jerome Laval, and his flight engineer Mark Hughes of Airtanker 64, airtanker pilot Tom Raider and his co-pilot Kris McAleer of Airtanker 02, airtanker pilot H. F. "Buzz" Schaffer, his late first officer Craig LaBare, and his flight engineer Tony Griffin of Airtanker 130, Battalion Chief Jim Stuller of the Dobbins Fire and Helitack Station in California, Captain Kevin V. Loomis and his Helitack Crew 408 (KCFD) from Keene, California, Air Attack Supervisor Fred Roach of the Kern County Fire Department command plane, foreman Dwayne Mortenson and his Helitack Crew 555 (BLM) of Bakersfield, California, Senior Pilot V. Lee Benson of the Air Operations of the Los Angeles County Fire Department, the crew of the Minden Air Attack Base (NDF) in Gardnerville-Carson City, Nevada, the Management Team of the Nevada Division of Forestry (Reno Airtanker Base), District Fire Chief Ed Lewis (Deer Park Airtanker Base), pilot Linda Scott of Leadplane 430 of Fresno, California, airtanker pilot Vito Orlandella of CDF Airtanker 100, and leadplane pilot Mike Lynn of Lancaster, California.

All of these firefighting aviators and many others have made it possible for me to study the technology and tactics of aerial firefighting in depth and experience them firsthand during numerous actions.

Naturally, "official" thanks go out to the forestry and forest fire agencies in the western states of the USA, above all the National Interagency Fire Center (NIFC) in Idaho, the U.S. Forest Service (USFS/USDA) and its many district offices, the Bureau of Land Management (BLM) and its regional facilities, the California Department of Forestry and Fire Protection (CDF/CAL FIRE), the Nevada Division of Forestry (NDF), and the California South Geographic Area Coordination Center (GACC) in Los Angeles, the Incident Command Center (ICC) in Bakersfield, California, and the Emergency Command Center (ECC) of the CDF and USFS in Fresno, California.

All these and many other agencies and offices have, completely without complications or problems, permitted me access to their individual regional and local facilities and assisted me with specialized information.

Last—but naturally not least—my thanks to all the photographers who have "supplied" me with their photographic material that I did not have available in my own archives. Most of them were friends and colleagues from fire departments in many non-American countries of the world and from the realm of international aerial firefighting, who made the "living visual proof" of their own activities available to me. My thanks go to Senior Inspector Roman Kostelnikov of the Russian Aerial Forest Fire Center Avialesookhrana in Pushinko near Moscow, Serghey Stelmakhovich of the Forest Science Institute in Krasnoyarsk, Russia, to the Director of the State Fire School in South Korea, Zyo Hionkoog, to Certified Engineer Sinisa Jembrih, member of the forest fire unit of the Professional Fire Department in Zagreb, Croatia, to Fragikas Dimitris of the Greek Firefighting Museum, Operation Manager Rob Erasmus of the Volunteer Wildfire Service in Capetown, South Africa, to District Supervisor Anthony Gray of the New South Wales Rural Fire Service in Australia, to Jake Oosthuizen of the Zululand Fire Protection Services in Africa, to Arnold Swart of the Chuma Safari and Conservation Services (Wildlife Services and Investigation) in Capetown, South Africa, to Fire and Environment Program Officer Dan Jamieson of Bright, Australia, to Jiri Sucharda of the Varnsdorf Fire Department in the Czech Republic and Ivo Mitacek, press spokesman of the Zlin professional firefighters, to the Federal Police aviators of Sankt Augustin, Germany, and to colleagues and friends Roman Vana of Olomouc, Czech Republic, Roland Oster (Aviation Picture) of Bad Kreuznach and Norbert Klekotko of Nuernberg, Germany, David Thomas of the USA, Pat McKelvey of Lewis and Clark County, Montana, Pete Dobbins of McKinney, Texas, Benoit Tenis of Canida and Richard Jud of the Swiss Embassy in Malaysia.

International manufacturers of aircraft and charter organizations have also supplied me with pictures of their flying products. Hearty thanks to, among others, the Director of Flight Operations, Christian Holm, of Neptune Aviation in Missoula, Montana, Vice President Kristin Edwards of Air Tractor Incorporation in Onley, Texas, Hugo Arceo of Air Tractor Europe in Sagunto-Valencia, Spain, Flight Operation Manager (CEO) Cedric Soriano of Air Attack Technologies of Marignane, France, and Operations Staff Manager Mason Bundschuh of Silver State Helicopters LLC in Las Vegas, Nevada.

Very hearty thanks to them all for all their assistance!

Wolfgang Jendsch
Author
Topic Editor, Fire Protection and Rescue Service
Member of the International Association of Wildland Fire (AWF), South Dakota.
Member of the California Fire Photographers' Association (CFPA), North Hollywood, California.